C0-AKS-018

THE VALUE OF CONVENIENCE

SUNY Series in Science, Technology, and Society
Sal Restivo, Editor

THE VALUE OF CONVENIENCE

A Genealogy of Technical Culture

THOMAS F. TIERNEY

State University
of New York
Press

Published by
State University of New York Press, Albany

Printed in the United States of America

For information, address State University of New York
Press, State University Plaza, Albany, NY 12246

Production by Susan Geraghty
Marketing by Lynne Lekakis

Library of Congress Cataloging-in-Publication Data

Tierney, Thomas F., 1957–
The value of convenience : a genealogy of technical
culture / Thomas F. Tierney.
p. cm. — (SUNY series in science, technology, and society)
Includes index.
ISBN 0–7914–1243–1 (HC) : $57.50. — ISBN 0–7914–1244–X (PB) :
$18.95.
1. Technology—Philosophy. I. Title. II. Series.
T14.T56 1993
601—dc20

91–42263
CIP

10 9 8 7 6 5 4 3 2 1

For Anne, Hannah, and Forrest

CONTENTS

ACKNOWLEDGMENTS

The perspective on technology that is presented in this text has taken shape over a rather long period. Some of those who were influential in first arousing and directing my critical interest in technological culture may be surprised to see where I have wound up after all these years. But thanks are due to Hwa Yol Jung and Gary Olson, professors at Moravian College who not only raised questions for me, but taught me to raise questions for myself. At the University of Massachusetts, William Connolly and Jean Elshtain exposed me to philosophical perspectives that broadened my field of view and enabled me to understand technology in the context of the body, necessity, and finitude. I am deeply indebted to them for this.

Jean Elshtain, Nicholas Xenos, Thomas Dumm, and Carlin Barton read early drafts of these chapters and offered questions and challenges that forced me to reconsider particular claims or interpretations, and thereby strengthened my argument. Of course, none of these readers (save one) bears any responsibility for the shortcomings or excesses of this text, and I assume that they themselves will still have problems with it. As they already know, this text was not written to please anyone.

I would like to add a special note of thanks to Thomas Dumm for the support he provided at all stages of this project, especially his early advice to wean myself from the writings of others and find my own voice. This gave me the confidence to abandon the plans I had for my argument and carry it in directions I had not foreseen. The conversations we had during the research and writing of this text always left me excited and anxious to continue, and were a catalyst that helped to make this publication possible. Dumm, therefore, cannot be completely absolved of responsibility for this text.

Since the issue of space is central to the perspective offered here, I would be remiss if I neglected to acknowledge the role that the Foster Homestead in North Calais, Vermont, played in shaping this text. The bulk of the research and writing took place at this location, and the place, as well as the people there, have left their mark on this work.

My deepest, most heartfelt thanks go to my wife, Anne, whose support and understanding enabled me to see this project through to its completion. She and our children, Hannah and Forrest, buoyed my spirits when I was struggling. And in a broader sense, I must also acknowledge my parents for the encouragement they have always given me. Finally, I must mention a special friend, Kelly, who was with me throughout this endeavor.

CHAPTER 1

Introduction

In this text, I will offer a critical interpretation of technological culture. While such interpretations were common in this century until about twenty years ago, today they appear outdated and outlandish, at least in their broad form. Narrower critiques, which focus on the technology of war, are still offered and received, but critiques which indict technological culture itself are no longer acceptable. The hope of both the East and the West is now centered on technological innovation and development, and 'less-developed' countries look to technology as the key to progress. Any challenge to this technological fetishism, therefore, is surely and sorely resented. Nevertheless, this text is, on the one hand, an attempt to revitalize the critical attitude of such thinkers as Jacques Ellul and Martin Heidegger, who viewed technological culture not only as a threat to alternative ways of life but also as a threat to the receptiveness to as yet unconceived possibilities. It is because of this menacing nature of technological culture that I seek to criticize and challenge it. But, on the other hand, this argument differs from many earlier critiques in several important respects.

To begin with, this text does not claim to reveal anything about the essence of technology, anything that is present in or underlies every manifestation of technology. Rather, what I offer here is nothing more than a perspective on technological culture. As a perspective, it is one among others, without any claim to special status because it has glimpsed something timeless in the phenomenon of technology.

To put it differently, this perspective treats technology as something which can be thought of along various lines, none of which is capable of revealing the heart of the matter of technolo-

---

Unless otherwise noted, all emphasis and parenthesis in quoted material is that of the author quoted, and any brackets are mine.

gy. It is only by approaching technology from various perspectives that one can begin to understand and, perhaps, resist it. And there is no reason for believing that after experiencing technology from various perspectives, one will be able to completely grasp it and utter a final word on the subject. So in regard to those interpretations which have been offered as revelations of the essence of technology, it is not so much that I find them wrong, but that I find they claim too much for their insights.

Another difference between this and many other perspectives on technology is that the one offered here does not trace the phenomenon of technique (Ellul), or the machine (Mumford), or *techne* (Heidegger), back to its origins. If the goal was to uncover the essence of technology, perhaps it would be necessary to follow the leaders back in their search for the original manifestations of technology. But even if one abandons the hope of glimpsing essences, there is still the temptation to extend one's perspective to include many of the historical developments of technology. Such a historical foundation provides a certain legitimacy to one's perspective, in the sense that one would appear to have a thorough understanding of the issue, and in the sense that one would be able to engage other leading perspectives (e.g., those of Jacques Ellul and Lewis Mumford) on many points.

Even if one could effectively borrow the legitimating form of essential, historical interpretations while renouncing their exaggerated claims, there is still reason for resisting the temptation to subsume the history of technology under one's perspective. By tying one's interpretation of modern technical culture to a long tradition of technical apparatuses, one recognizes the important innovations in technological development, but at risk of losing sight of the web of relations, or better, the lines of power, through which technology flows in modernity. And it is through such an ensemble of lines that technology helps to form and shape the modern self. Since the primary concern of this text is the fetishistic attitude of the modern self toward technology, I will focus only on modern technology, and even then the concern will be primarily with the relation between people and technical culture, and not simply with the features of technical apparatuses.

It must be emphasized that this imposition of limits on the historical treatment of technology is not offered as a methodological principle which is to be universally applied. I am not making the claim that modernity can be understood only on its own terms, that only by focusing on the modern can one understand modernity. On the contrary, this text will develop a broad historical perspective, but it is one that does not take the phenomenon of technology as its central theme. Instead, modern technology will be portrayed as an element of a different historical line, one which reveals aspects of technology often overlooked by histories of technical development.

In its treatment of modern technology, this perspective differs in a third way from many other perspectives on technology. This difference lies in what I, but not they, would describe as the "line of attack." Many interpreters of modern technology focus on the way in which technology expands and invades every facet of nature and/or society, establishing an order throughout. I have in mind here interpreters such as Ellul and Heidegger.[1] There is no doubt that technology does expand in such a manner and that it does tend to engulf not only nature, but all human activities as well. But by focusing on this expansion, and mapping out the advances of technology, one does little to foster resistance to the power of technology. Indeed, Ellul's monolithic portrayal of modernity in *The Technological Society* leaves virtually no room for resistance. But there is resistance to technical culture.

A paradoxical example of this resistance is the rise of Islamic fundamentalism, which rejects the technological culture of the West.[2] In its resistance to this culture, fundamentalism has indeed employed certain military techniques and apparatuses from the West and has also developed terroristic techniques of its own, but the point is that this technology is directed against the ever-expanding technical culture in order to resist it. And even within technical culture itself there are subterranean economies which lie beyond the control of the economic techniques of the state, acts of sabotage and protests which are intended to thwart the deployment of new military and nuclear-power technology and, more recently, living wills and suicide machines which resist modern medical technology. Without getting into the merits of any of these forms of resistance, the point is simply that technical culture is not nearly so tightly ordered or efficient as some have portrayed it. Resistance, however effective, occurs at various levels.

In this attempt to challenge technical culture, I will not focus on the imperialistic character of modern technology. This is not to deny that it might be worthwhile to draw a map which complements the one of technological expansionism, and points out the various ways in which technology is resisted as it expands in society and nature.[3] But the resistance which this argument strives to incite is found in a different area, or on a different level, and therefore requires a different approach. Instead of focusing on the way in which technical culture expands, this text is concerned with the way in which it becomes narrow and pointed, the way it penetrates and shapes modern individuals and renders them techno-fetishists. In other words, the concern here is with the way in which technology affects the values of individuals.

Two basic questions can be asked at this level. First, what is the value of technology to modern individuals? And second, why do they hold this value in such high esteem that, even when faced with technological dangers and dilemmas, they hope for solutions that will enable them to maintain and develop technical culture? Before I begin to answer these questions, however, there are a few points that must be made about inquiries carried out at the level of values.

The first of these points is that the interpretation of technological culture from the perspective of values does not constitute a novel approach to this question. Early in the twentieth century Max Scheler pointed out that, despite its claim of value-neutrality, modern science (as well as its technological application) was guided by a particular value—namely, the domination of nature.[4] It is worthwhile at this point to briefly examine Scheler's insight into technical culture, both because there are certain similarities between Scheler's approach to the question of technology and mine, and because Scheler's insights were developed by later theorists in a manner I will assiduously avoid. But even beyond these reasons for looking at Scheler's thoughts on technology, the value Scheler ultimately identified as dominant in technical culture is a complement to the one I will emphasize.

To begin with, Scheler's approach to understanding a given culture consciously focused on values. It was not just that scientific knowledge was not value-free; for Scheler, no form of knowledge or action could be. Echoing Nietzsche's claim that "the question of values is more *fundamental* than the question of certainty,"[5] Scheler wrote that "all perceptions and thoughts, with regard to the laws governing the *selection* of their *possible* objects, and, not any less fundamental, all our actions, are rooted in the *conditions of valuation and drive-life*."[6]

Nietzsche's "profound influence"[7] upon Scheler, however, extended far beyond the latter's general recognition of the primacy of values and valuation. Scheler also shared Nietzsche's critical perspective toward the dominant values of a culture and relied heavily on Nietzsche's genealogy of Christian morality for insights into the values of modernity. While Nietzsche identified the resentment which the weaker, priestly caste felt toward the stronger, aristocratic types as the primary motivation for Christian morality and its modern variants,[8] Scheler also pointed to resentment as the primary motive beneath modern values.

The shift in values which marked the break between the medieval and modern periods, according to Scheler, was the substitution of the value of utility for the spiritual values which were predominant in medieval culture. And this transformation was motivated by the resentment that the bourgeoisie felt toward the values of the more spiritual, aristocratic types. As Scheler put it, utilitarianism was the "chief manifestation of the *ressentiment* slave revolt in modern morality."[9]

Later in his career, Scheler changed his mind about the dominant value of modernity, especially in regard to the technological prowess of this age.

> The basic value that guides modern technology is not the invention of economical or 'useful' machines.... It aims at something much higher.... It is the idea and value of *human* power and human *freedom vis-a-vis* nature that ensouled the great centuries of 'inventions and discoveries'—by no means just an idea of utility. It concerns itself with the *power* drive, its growing *predominance* over nature *before* all other drives.[10]

Scheler pointed out that in the feudal period, the power-drive had been directed at the domination of other persons, but in the modern period, the domination of nature was the object of the power-drive; he called this modern drive "the will to control nature."[11] Some contemporary thinkers have further developed Scheler's insight into modernity's drive to dominate nature, but before turning to this development I must point out other similarities between Scheler's perspective on technological culture and the one to be developed in this text.

Scheler and I are both heavily indebted to Nietzsche for the conceptual schemes that we develop. Following Nietzsche's insights into the primacy of values and valuation, Scheler uncovered the values which underlie the professed neutrality (i.e., value-freedom) of modern science and technology. And, of course, the larger historical framework into which Scheler fits the modern ethos is a Nietzschean one. I, too, take my clues about the value of technology from Nietzsche, although the value I will emphasize is neither the value of utility nor the domination of nature. I take my lead from Zarathustra, who said upon his return to others and their cities:

> I go among this people and keep my eyes open: they have become *smaller* and are becoming ever smaller: *and their doctrine of happiness and virtue is the cause.*
>
> For they are modest even in virtue—for they want ease. But only a modest virtue is compatible with ease.[12]

This desire for ease will be the primary focus of this text. For etymological reasons which will be discussed in the following chapter, I choose to call the object of this desire "convenience" rather than ease. In any case, the main contention of this argument will be that the value of technology in modernity is centered on technology's ability to provide convenience. The aim of my text, however, is not to lament the smallness or mediocrity of modern individuals and their virtues. It is rather to throw some light on, and thereby loosen, the hold which technology has on modernity. The desire for convenience seems to be an integral part of that hold—that is, an integral part of the modern self.

The larger historical trend into which I will ultimately fit my discussion of convenience is also a trend which Nietzsche traced, and in this, too, my argument bears a certain resemblance to Scheler's. While Scheler turned to the first essay of *The Genealogy of Morals* for his historical perspective, I will rely on the third essay, in which Nietzsche outlines the history of the ascetic ideal.

Although a claim that technical culture somehow fits in with the history of asceticism may seem incomprehensible at this point, this connection should become clearer once the idea of convenience has been fleshed out.

One more similarity between Scheler and myself must be noted, and this similarity has to do with the manner in which we approach the values of the technical age. In identifying utility and, later, the will to control nature as the primary values of this age, Scheler's aim was to criticize those values by showing how they emerged from a certain baseness. In this critical endeavor, Scheler can be thought of as a genealogist, at least in the sense of genealogy expressed by Gilles Deleuze:

> Genealogy means both the value of origin and the origin of values. Genealogy is as opposed to absolute values as it is to relative or utilitarian ones. Genealogy signifies the differential element of values from which their value itself derives. Genealogy thus means origin or birth, but also difference or distance in the origin. Genealogy means nobility and baseness, nobility and vulgarity, nobility and decadence in the origin. The noble and the vulgar, the high and the low—this is the truly genealogical and critical element.[13]

I must emphasize that the claim being made here is not that Scheler was a thoroughgoing genealogist. Despite Nietzsche's influence, Scheler did attempt to construct an absolute hierarchy of values,[14]and he also tried to rescue the essence of Christianity from Nietzsche's attack.[15] But in regard to his interpretation of the underlying value of modernity, Scheler was doing genealogy. He treated neither utility nor the will to control nature as the logical outcome of historical progress or as a value grounded in some fact of human existence. Rather, these values were regarded as the outcome of certain shifts in relations of force, as the outcome of a reversal in the struggle between the noble and the base. In my treatment of the value of convenience, I share this genealogical attitude toward values, which treats them as the signs of a struggle, and I also attempt to criticize and reevaluate this particular value.

In a sense, Scheler and I offer complementary genealogies of modern values. The value upon which Scheler focused—the domination of nature—has been the value which guides the cutting edge of technology; it is the value pursued by the leaders of technological progress, the scientists and technicians. The value of convenience, on the other hand, is the value of the masses, of those who consume the products of technical culture.[16] But, as will become apparent, the value of convenience (in an extended sense of the word) has come to lead certain aspects of technological innovation and development as well. For now, however, all I want to do is point out the complementarity of Scheler's genealogical project and the one offered here.

While Scheler's genealogical impulse marks a particular affinity between our perspectives on modernity, this same impulse distinguishes Scheler from certain others who have developed his insight into the domination of nature. I have in mind here theorists such as Max Horkheimer, Herbert Marcuse, and William Leiss, all of whom can be considered critical theorists in the sense first articulated by Horkheimer.[17] These thinkers coupled Scheler's insight with the dialectic, thereby eliminating "the truly genealogical and critical element," or stated differently, the Nietzschean element, of Scheler's thought. Since Scheler is valuable to me primarily for that Nietzschean element, I must briefly examine this coupling of the will to control nature and the dialectic. Such an examination will reveal the grounds for my avoidance in this text of any dialectical interpretation of the value of convenience. It will also lay the foundation for the claim which will be made later that critical theory (Marcuse, in particular), rather than pulling in the reins on technology actually spurs it on into new areas of development.

Critical theorists such as those mentioned above accept, tacitly or explicitly, Scheler's claim that science is not value-free, but rather serves the value of dominating nature.[18] But these theorists point out a shortcoming of Scheler's thought: he neglected to take into account the social context in which such domination occurs. Consequently, Scheler remained blind to the fact that under existing social conditions of injustice and inequality, the scientific domination of nature results in the ever-increasing domination of people through—and by—technology. In the words of William Leiss, "Advances in technology clearly enhance the power of ruling groups within societies and in the relations among nations;

and as long as there are wide disparities in the distribution of power among individuals, social groups, and states, technology will function as an instrument of domination"[19]—the domination of people, that is.

It is here that the dialectic is grafted onto Scheler's thought. The will to dominate nature is rendered contradictory, irrational by this negativity of social injustice and inequality. And through the elimination, or negation, of this negative social atmosphere the will to dominate nature can be rendered rational, and technology will finally be able to fulfill its original goal of promoting human freedom and security.

Through the use of the dialectic, therefore, critical theory has been able to salvage the will to control nature. The irrational, dangerous trajectory of technology in modernity stems not from the value of dominating nature, but from the injustice of advanced industrial society. Critical thought, consequently, must work toward the elimination of relations of domination and subordination among people. As Marcuse put it, this elimination is "the only truly revolutionary exigency, and the event that would validate the achievements of industrial civilization."[20] It would also validate the will to control nature, and the critical theorists mentioned here do indeed expect that any just society of the future would have to carry on the conquest of nature.[21]

This salvaging which is accomplished by critical theory's use of the dialectic is precisely what makes it unacceptable to me. Instead of carrying out a ruthless criticism of what Scheler identified as the will to control nature, critical theory ends up making it acceptable, rational. This reveals the extent of the dialectic's critical capacities. It is able to turn things on their heads, transforming the decadent will to dominate nature into a noble goal to be pursued into the future, but this dialectic is not capable of cutting off the head of such a decadent value and being done with it. Given the context in which this discussion of the dialectic has emerged, Gilles Deleuze's judgement of it seems particularly appropriate:

> *It* [the dialectic] *is reactive forces that express themselves in opposition, the will to nothingness that expresses itself in the labour of the negative.* The dialectic is the natural ideology of *ressentiment* and bad conscience. It is thought in the perspec-

> tive of nihilism and from the standpoint of reactive forces... powerless to create new ways of thinking and feeling.[22]

From my perspective, however, the most objectionable feature of the dialectic is not so much its "*ressentiment*," which is revealed in its formal properties of negation and reaction, but its "bad conscience"—that is, its inability to forget, to let go of bad memories and nihilistic values. It is primarily for this reason that my treatment of the value of convenience will not be dialectical. I will not portray convenience as a certain negativity which has derailed the rational progress of science and technology, and which must be negated so that technical culture can become noncontradictory and capable of fulfilling its promise (threat). My goal is not to save technical culture, but to undermine it. I will also not portray convenience as an inherently noble value which has itself been sidetracked by some social negativity, such as economic and political injustice, the elimination of which would allow convenience to flower in an environment of reason and freedom. From my perspective, the desire for convenience is a weed, not a flower, and my objective is to uproot it.

While the perspective that I am developing may appear extreme (with its images of decapitations and vegicide), and perhaps unreasonable (in its implied belief that a value which has been carried along and fostered by modern tradition can actually be uprooted), such excesses seem to me justified by those very considerations which would give rise to these objections. Because it is so deeply ingrained in modern culture, the value of convenience can only be challenged by an aggressive attack.[23] A reckless, all-out effort is required just to create the space from which this value can be challenged.

Additional considerations justify the excesses of this genealogy of convenience, but these have less to do with the traditional inertia of convenience than with the broader tradition of liberal individualism. Any inquiry into values faces resistance from this liberal tradition, which recognizes at the core of the individual a private realm which lies beyond the reach of social and cultural forces.[24] This private realm is one of beliefs, intentions, desires, and—most importantly for this text—values. Although liberalism's claim of privacy in this sphere was challenged by nineteenth-century social theorists such as Hegel and Marx, it still exerts enormous influence on the self-understanding of modern individuals and is tightly bound up with their claim to freedom. Stuart Hampshire articulates this influence when he writes, "The man who is comparatively free in the conduct of his life is active in the adoption of his own attitudes and of his own way of life; his decisions and intentions are the best guide to his future action; and just this is the significance of calling him free."[25] It is to be expected, therefore, that an argument such as mine, which claims that a certain value is not freely chosen by individuals, but is demanded by various facets of the technological order of modernity, will be met with a degree of self-preserving (in a very literal sense) denial.

This liberal resistance to inquiries into values is compounded in the case of my argument because that argument is an invasion of privacy in a second sense, one which is derived in part from the classical Greek conception of privacy. For the ancient Greeks, the private realm was not located within the individual, as a sphere of beliefs, values, and intentions, but rather, it was located in the household. My inquiry into the value of convenience will begin in the modern household, which, I will argue, still retains elements of the classical conception of privacy. I will begin in the household because it is there that convenience reigns, there that the self is shaped by the demands of the technological order, and there that individuals 'buy into' technical culture.

My argument challenges at once the privacy of the individual and the privacy of the household (although these are not unrelated spheres). Because my text is an invasion of privacy, or a trespass, in this double sense, it is bound to face resistance. To some extent, therefore, the success of this text can be measured in the amount of resistance that it evokes. But the most serious threat posed to individuals today does not come from arguments that challenge the privacy of the realm of beliefs, values, and desires, but rather from unchallenged forces that penetrate that sphere. The value of convenience is one such force.

The course of this genealogy of convenience begins with an examination of the modern household, in the context of Hannah Arendt's interpretation of that household in *The Human Condition*. My purpose in the second chapter is to challenge Arendt's claim that modernity is characterized by a "reverence" for the body. Ultimately, I will argue that the consumption of convenience in modernity reflects a certain contempt for the body and the limits it imposes, and for those readers familiar with Arendt's argument, it should be apparent that there is a clash between her interpretation of modernity and mine.

After discussing Arendt's argument, I turn to some contemporary Marxist interpretations of modern consumption practices. In part, my aim here is to acknowledge that these Marxists have moved beyond the rigid structuralism of earlier generations of Marxist scholars, but my concern also is to indicate limitations of this Marxist perspective on consumption. Ultimately, these writers interpret modern consumption practices as being determined by the demands of the production process, and this blinds them to other important influences on consumption practices, especially in the case of the United States, which most of these writers accept as the epitome of modernity.

I then offer a very different interpretation of American consumption standards, one which challenges the interpretation of the Marxists I criticize, but which is nonetheless based on a particular insight Marx had concerning the uniqueness of the United States. Marx realized that the spatial dimensions of the United States posed serious challenges to capitalism, even if he did not recognize the impact that unlimited space would have on modern consumption practices. Capitalism's response to the problem posed by unlimited space, I argue, played an important role in establishing the value of convenience as the driving force behind modern attitudes toward technology.

This genealogy of convenience, however, is not simply or purely materialist. Alongside the spatial situation in the United States, other factors played equally important roles in the emergence of convenience as a primary value in modernity. The decline in religious belief commonly associated with modernity is one of these factors, and I focus on this dimension of the technological question late in the text. I approach this subject in the context of Max Weber's controversial argument in *The Protestant Ethic* and expand that argument with the help of Nietzsche's insights into Protestantism and asceticism. Ultimately, I will claim that the fetishistic attitudes toward technology and the rampant consumption of 'conveniences' which characterize modernity are a form of asceticism. In one of the last chapters of the text, I uncover evidence of this ascetic dimension of modernity in several modern political thinkers, ranging from liberals to radicals.

The thread which runs throughout this wide-ranging array of evidence, I should perhaps reiterate, is the value of convenience. Although this value is not usually the object of discussion or reflection, it nevertheless holds a highly esteemed position today and guides the consumption choices of individuals in modern technical culture. What I hope to accomplish by following these very different lines of approach to this value is to throw convenience into relief, to make it noticeable, questionable, and hopefully, challengeable.

CHAPTER 2

Arendt,[1] the Household, and Convenience

In *The Human Condition*, Hannah Arendt points out the difference between the classical Greek and the modern conceptions of privacy. In ancient Greece, the realm of privacy was not found within the individual or the subject, as it is in modernity. Instead, privacy was recognized as something inherent in the household. The difference between these conceptions of privacy was not limited to the location of the private, however. For the Greeks, the household was considered private not because it was the realm of beliefs, desires, and values, but because it was the realm in which biological necessity prevailed. In the ancient household occurred the production and consumption required to sustain life,[2] and in the performance of this necessary activity the Greeks recognized the fundamental similarity between themselves and other animals. Certain routines were imposed upon all animals, including the Greeks, as a consequence of their embodiment. However, the Greeks distinguished the human animal from other animals precisely by its ability to free itself from the routines imposed by its body, and to undertake meaningful, unnecessary activity. That is, the Greeks distinguished themselves as humans by their ability to move beyond the concerns that serve the maintenance of biological, physical life, and to undertake inquiries about the ultimate purpose or ends of life. For Aristotle, the unique thing about humans was not simply their capacity for rational speech, but their capacity to rationally discuss the proper ends of life set them apart from other animals.[3] It was through such discussions, and the attempt to live according to the knowledge revealed by them, that people became fully human.

In ancient Greece, the freedom from necessity which provided the opportunity for such discussion was attained by certain adult males through the practice of slavery and the rigid differentiation of the sexes. Women and slaves performed most of the necessary activity in the household, while free, adult males attained human status through their participation in the discussions, debates, and decisions of the *polis,* the public realm. The Greek household, therefore, was private in the sense that those whose roles were limited to performing its necessary activity were deprived of the opportunity of being fully human. As Arendt puts it: "In ancient feeling the privative trait of privacy, indicated in the word itself, was all-important; it meant literally a state of being deprived of something, and even of the highest and most human of man's capacities."[4]

Arendt argues that this privative dimension of privacy has been lost to modernity, that the private is no longer the realm of subhuman, slavish activity. On the contrary, modernity's conception of privacy is closely linked to its ideal of freedom. The "sphere of intimacy" (as Arendt calls the private realm of modernity[5]) is a haven, not a hellhole. Arendt's perspective on modernity, of course, encompasses much more than this shift in the status of the private from the position of unfreedom in ancient Greece to that of freedom in modernity. She identifies several other transformations that have occurred alongside this shift.

One of those related transformations which is particularly important for this text (and also particularly disturbing to Arendt) is the severance of the ancient link between necessity and the household. The demands of the body are no longer satisfied within the private household but have been swept out into the open. The activity necessary to sustain life is now performed as social, not private, activity. For Arendt, "Society is the form in which the fact of mutual dependence for the sake of life and nothing else assumes public significance and where the activities connected with sheer survival are permitted to appear in public."[6] In other words, society is a form of public household.

In making this claim, Arendt is not referring primarily to the activity of the liberal or socialist state, which provides certain goods and services to individuals. Rather, the economy is what Arendt has in mind when she refers to the social. It is as participants in the economy, as jobholders, that modern individuals appear in public. The predominant concern for members of mass society lies in satisfying the demands of the life processes for themselves and their dependents.[7] 'Productive members of society,' as opposed to ancient citizens, are not concerned with the ultimate ends of human life, but rather with simply making a living. This is all there is to the public activity of the member of society.

Arendt has a certain difficulty with the emergence of this social realm (which she dates as beginning around the sixteenth century[8]). It appears that the source of this difficulty lies with the effect the development of this realm has had on public activity. When the guiding force in public life is the attainment of the wherewithal to satisfy the life processes, there is no longer any room or time left to pursue the unnecessary—and thereby human—goals of the Greek *polis*. The Greek desire to attain some measure of immortality through the public presentation of great works, words, or deeds[9] finds no quarter in an age where mortal, bodily considerations prevail. In such an age, political life is reduced to bureaucratic administration, a sort of public housekeeping which tries to organize social laboring.[10]

It seems, however, that there is more to Arendt's dissatisfaction with the modern social realm than its devaluation of heroic political action. There is something else about modernity which bothers Arendt, and this is indirectly indicated by her treatment of Christianity. As Arendt points out, the decline of the ancient public realm began with Socratic philosophy, which identified the contemplative life as superior to an active life in the *polis*.[11] This denigration of public activity was maintained and developed by Plato, Aristotle, and the Stoics, but it was Christianity, Arendt claims, which transformed the contemplative, apolitical life of the philosophers into "a right of all."[12] Indeed, Arendt writes that the fall of the Roman Empire and the diffusion of the Christian gospel of an eternal afterlife together made any striving for an earthly immortality futile and unnecessary:

> And they succeeded so well in making the *vita activa* and the *bios politikos* the handmaidens of contemplation that not even the rise of the secular in the modern age and the concomitant reversal of the traditional hierarchy between action and contemplation sufficed to save from oblivion the striving for immortality which originally had been the spring and center of the *vita activa*.[13]

If Arendt's only gripe against modernity were its lack of a sphere of public activity (in the Greek sense), one would expect that she would find the denial of earthly immortality—the spring and center of Greek political life—to be the most significant feature of Christianity. But this is not the case. Rather, she identifies the revaluation of life itself as "the most important reversal with which Christianity had broken into the ancient world."[14] As has already been indicated, the Greeks held biological life and its demands to be something less than human. Arendt claims that Christianity reversed that Greek attitude toward physical life and elevated life to the level of the sacred.[15] By this, Arendt means that Christianity viewed life on earth, or mortal life, as essential to the attainment of the eternal life offered by Christ. Only through life on earth could one enter heaven. Arendt goes so far as to claim that, according to Christianity, "to stay alive at all costs had become a holy duty."[16]

The difference Arendt finds between the Greek and Christian attitudes toward life is revealed by comparing their different attitudes toward suicide. Arendt points out that part of the Greek contempt for the slave was based on the slave's choice of a life of slavery over death.[17] By refusing to commit suicide rather than live as a slave, the slave was repulsive to the Greeks. For one about to become enslaved, the Greeks found death to be a more noble choice than life. In support of her claim that Christianity reversed the Greek attitude toward life, Arendt cites the Christian refusal to bury on blessed ground those who commit suicide.[18] Those who choose to kill themselves, instead of continuing to live even the most wretched existence, cannot enter heaven and are denied eternal life.

One possible reason why this Christian elevation of life is more important for Arendt than its denial of earthly immortality is that this reverent attitude toward life has become one of the central features of modernity. The Christian belief in an otherworldly immortality, on the other hand, has been abandoned. Christianity's "fundamental belief in the sacredness of life has survived, and has even remained completely unshaken by, secularization and the general decline of the Christian faith."[19] The form which this reverence for life takes in modernity is, according to Arendt, the overarching concern for the production process or the economy. Through the productive activity of humans the life of the species is preserved.

Although concern for the life of the species rather than the individual is a significant (if overstated) difference between modern society and the Christian community, reverence for life runs through both ages. In this sense, Arendt sees Marx as an unwitting smuggler of Christian attitudes into modernity. But the fact that the sacredness of life exerts its influence on modernity is not a sufficient explanation for Arendt's ranking it as the *most* important effect of Christianity. There is no reason for treating the longevity of this influence as an indication of its importance.

Another possible explanation for her ranking can be found in the detrimental effect which the modern form of the elevation of life has had on public life. After all, Arendt explicitly claims that society, in which the life of the species is preserved, continually encroaches upon political life, further decreasing the possibility for public action.[20] But this explanation is also insufficient. For if the standard for her ranking is the effect which Christianity has had on public life, it would appear that the belief in an otherworldly afterlife, which cast the desire for earthly immortality into oblivion, would be more important than the revaluation of life.

A more sufficient explanation of Arendt's ranking of the effects of Christianity is available, but this explanation is grounded less upon Arendt's explicit concern with the decline of politics and public life than with an underlying, muted concern of hers. This concern is for the loss of the ancient private realm, in and of itself, and not because of the effect this loss has had on the public sphere. Arendt thinks that certain human activities—those associated with the body and the life processes—should not be seen, but rather that they should be kept from public view. In ancient Greece, these activities were hidden in the household. And the reason these activities were—and from Arendt's perspective should be—hidden is that they are not worthy of public exposure.[21]

In other words, there seems to be in Arendt's thought a certain contempt for the body and the life processes, a contempt she recognizes and admires (and, I will argue shortly, overemphasizes) in the Greeks. The Greeks considered the slave contemptible not because he hindered public life (indeed, such a public life was possible largely because of the slave), but because he displayed a certain baseness in clinging to life above all else. In his defense of slavery in *The Politics*, Aristotle cites as natural slaves those "whose condition is such that their function is the use of their bodies and nothing better can be expected of them."[22] The body is inferior to the mind, and those whose bodies are stronger than their minds are inferior to mentally developed and active individuals.

Arendt's ranking of the effects of Christianity, I am arguing, is similarly based on the attitude that the body and the life processes are base. By making physical, mortal life an essential element in the attainment of an eternal afterlife, Christianity weakened the stigma that the Greeks had attached to the life processes.[23] Household activity was no longer considered subhuman, but was raised instead to the level of a divinely sanctioned element of human activity.[24] Given the underlying contempt which I suggest Arendt has for such activity, it is understandable that she would claim that the Christian revaluation of life is the most important reversal accomplished by Christianity. It marked the beginning of the end of the ancient realm of privacy.

This contempt for the life processes is what I was referring to earlier when I claimed that there is more to Arendt's dissatisfaction with society than her concern for its lack of heroic public activity. The emergence of society, which is the modern form of the sacredness of life, brought about the complete collapse of the ancient private realm. Even during the long reign of Christianity, the demands of the body were still satisfied within the household; it was only the status of the household and its activity which had changed with Christianity.[25] In society, however, the life processes are no longer hidden within the household, but occur out in the open, in public. And although Arendt writes that "it is striking that from the beginning of history to our own time it has always been the bodily part of human existence that needed to be hidden in privacy, all things connected with the necessity of the life process itself,"[26] my claim is that it is disturbing, perhaps disgusting, to her that this is no longer the case. Such necessary activity, argues Arendt, should by its very nature, be hidden from public view.

I will eventually argue that Arendt's attitude toward the body and necessity is a quintessentially modern characteristic, and I maintain this despite Arendt's self-conscious stance as one repulsed by modernity. But before I move on to an examination of some additional consequences of Arendt's contempt for the body, I should briefly mention and respond to an objection that might be raised concerning the existence of this contempt. Some commentators on Arendt's thought have pointed out the important role which "natality" plays in her thought.[27] In *The Human Condition*, Arendt claims that "[t]he miracle that saves the world, the realm of human affairs, from its normal, 'natural' ruin is ultimately the fact of natality, in which the faculty of action is ontologically rooted."[28] While this reference to natality may seem to belie my claim about Arendt's contempt for the body, in that Arendt places such great importance on this bodily function, I think that her use of natality instead supports my claim.

First, the example of natality which Arendt cites is the birth of Jesus Christ. She could hardly have chosen a less typical example of human natality. The birth of Christ was not the result of the sexual union of a man and a woman, but was instead the result of immaculate conception. Furthermore, Christ did not die the typical death of an embodied human. He died, and then rose from the dead. The images of Immaculate Mary, Virgin Mother, and the risen Christ are at best sanitized versions of the human body. At worst, they are denials of the body.

More important than Arendt's choice of an example of natality, however, is the use to which she puts this notion. The last clause of the previous quote, which concerns "the faculty of action," reveals the point which Arendt is trying to make. In the paragraph preceding the one from which the quote was taken, Arendt discusses the faculty of action:

> If left to themselves, human affairs can only follow the law of mortality, which is the most certain and the only reliable law of a life spent between birth and death. It is the faculty of action that interferes with this law because it interrupts the inexorable automatic course of daily life, which in its turn, as we saw, interrupted and interfered with the cycle of the biological life process.[29]

The faculty of action, in other words, interrupts the daily routine of laboring activity (the inexorable automatic course of daily life), which is the human response to the law of mortality. Action, in this sense, is a way of breaking out of the routines imposed by human mortality. Therefore, natality, as the ontological ground in which action is rooted, should not be interpreted as an Arendtian celebration or affirmation of the body and the life processes. On the contrary, as Arendt uses the term, natality is the source of hope that people can transcend the limits imposed by their embodiment and accomplish something immortal. As such, the concept of natality supports rather than weakens my claim about Arendt's contempt for the body.

Although this contempt for the body is an underlying theme of *The Human Condition*, it has a profound influence on Arendt's text. Her interpretations of modernity and Christianity, not to mention her interpretation of classical Greece, are all shaped by this attitude toward the body. More to the point, I think that this contempt leads Arendt to misinterpret each of these cultural periods. From my perspective, neither Christianity nor modernity holds life to be as sacred as Arendt claims, nor did the Greeks find the body to be thoroughly contemptible. In each of these cases, Arendt's interpretation seems to me skewed by her overreaction to the realm of the body and its needs.

Since the primary focus of this text is the modern, technological age, Arendt's interpretation of modernity is obviously of concern here, but just as important is Arendt's interpretation of Christianity. This is because our different perspectives on modernity are greatly influenced by our different interpretations of Christianity. Contrary to the postmodern infatuation with discontinuity and rupture, Arendt and I both identify lines of continuity between Christianity and modernity, although the lines we identify are remarkably different.

In emphasizing Christianity's treatment of physical, biological life as a prerequisite for entrance into the heavenly kingdom, Arendt ignores what might be considered the other side of the Christian attitude toward life. This other side, which is grounded in the Book of Genesis account of humanity's fall from grace, treats the "law of mortality" and the demands of the life processes as punishment for the original sin. Although there were at first different Christian interpretations of the relevant sections of Genesis,[30] the reading offered by Augustine became the official teaching of the Roman Catholic church.

According to Augustine's interpretation of Genesis, mortal life with all its toil and trouble was the punishment all must suffer as a result of Adam's sin. Initially, God had created for man a garden in which he would live, and out of the ground of that garden "the Lord God caused to grow every tree that is pleasing to the sight and good for food; the tree of life also in the midst of the garden, and the tree of the knowledge of good and evil" (Genesis 2:9). (All biblical quotations are taken from *New American Standard Bible,* Reference Edition [Chicago: Moody Press, 1975].) Adam was forbidden to eat of this last tree, the tree of knowledge of good and evil. God warned him that "in the day that you eat from it you shall surely die" (Genesis 2:17). It was with the fruit of this tree that the serpent tempted woman, telling her that "in the day you eat from it your eyes will be opened, and you will be like God, knowing good and evil" (Genesis 3:5).

Adam and Eve, of course, ate of this tree and were consequently expelled from the garden by God. They did not immediately die, however, as one might expect from the warning that God had given them. The point of the expulsion was to prohibit humans from eating of the tree of life, and thereby to deny them the possibility of attaining everlasting life. It was in this sense of losing eternal life that Adam died. As God said to Adam upon learning of the sin, "'Behold, the man has become like one of Us, knowing good and evil; and now, lest he stretch out his hand, and take also from the tree of life, and eat, and live forever'—therefore the LORD God sent him out from the garden of Eden" (Genesis 3:22–23). And to keep him out, God stationed a band of angels and "the flaming sword which turned every direction, to guard the way to the tree of life" (Genesis 3:24).

Mortality, therefore, was not part of the original human condition but the result of the first sin. But there was more to the punishment than the denial of everlasting life; God also condemned humanity, in its mortality, to a life of increased toil and labor. Before the fall, bodily needs were easily satisfied. The fruit of the trees satisfied human hunger and water flowed throughout the garden to satisfy thirst and to water the trees (Genesis 2:10).

But after the sin, God said to Adam, "Cursed is the ground because of you; in toil you shall eat of it all the days of your life. Both thorns and thistles it shall grow for you; and you shall eat the plants of the field; by the sweat of your face you shall eat bread, till you return to the ground" (Genesis 3:17–19).

There was also a special facet of the punishment, a particularly cruel one, that God directed toward woman. Because she succumbed to the serpent's temptation and led man astray, God said to the woman, "I will greatly multiply your pain in childbirth, in pain you shall bring forth children; yet your desire shall be for your husband, and he shall rule over you" (Genesis 3:16).

I should emphasize that I am not claiming that the Catholic Church taught that prior to the fall there was no labor in either the procreative or the toilsome sense of the word. Arendt forcefully criticizes those who make such claims and points out that the punishment consisted in making labor (in both senses) more burdensome and painful.[31] I agree with Arendt's interpretation of the Old Testament—on this point, at least: before the original sin, there was some pain in childbirth, and the fruit trees which God provided did require some cultivation by man.[32]

But although I agree with Arendt that labor was not created, merely intensified, by the punishment for original sin, I disagree with her concomitant claim that death, or human mortality, was not a result of the fall from grace. In support of this last point, Arendt simply states that "nowhere in the Old Testament is death 'the wage of sin.'"[33] There is no arguing with this; the phrase Arendt quotes is from Paul's epistle to the Romans (6:23). But this simple statement of Arendt's does little to support her claim that Christianity held life to be sacred. From Augustine forward, the Christian interpretation of Genesis held that mortality was indeed the wage of sin. And this claim was based in large part on the last verses of the third chapter of the Book of Genesis, some of which have been quoted here. The point of these verses, as interpreted by Augustine, Luther, and Calvin (I will get to Luther's and Calvin's interpretations in Chapter 5) is that God expelled humans from the garden and blocked their return to the tree of life to prevent them from eating of this tree and thereby attaining eternal life.

Before it gets lost in this biblical quibbling, the point which I am trying to make is that, contrary to Arendt's claim, Christianity did not simply hold mortal life and the life processes to be sacred,[34] but rather it viewed human mortality and the burdensome, painful nature of the life processes as the results of original sin. This is not to say that Christianity held life and biological necessity in the same contempt which Arendt found among the Greeks, but that the Catholic Church and the Reformers accepted necessity and mortality in the spirit of guilt for the primal sin. As descendants of Adam, all humans share in his guilt and punishment.

In the New Testament, the figure of Christ comes to redeem humanity from this guilt and to offer salvation. And while it is true that this redemption was accomplished precisely by God's becoming human, and by suffering, dying, and being buried,[35] there remains in Christianity a certain ambivalence about mortal life that Arendt misses. On the one hand, it is only by living according to the example of Christ's life on earth and then dying that one can be redeemed and attain everlasting life; on the other hand, death is recognized by Christianity as a punishment for sin. This ambivalence is expressed quite clearly by Augustine:

> Undoubtedly, death is the penalty of all who come to birth on earth as descendants of the first man; nevertheless, if the penalty is paid in the name of justice and piety, it becomes a new birth in heaven. Although death is the punishment of sin, sometimes it secures for the soul a grace that is a security against all punishment for sin.[36]

As for Arendt's claim that "to stay alive at all costs had become a holy duty"[37] under Christianity, the gospels of Christ's apostles, not to mention the deaths of the martyrs, clearly refute this claim. According to Mark, Christ proclaimed to his followers, "For whoever wishes to save his life shall lose it; and whoever loses his life for my sake and the gospel's shall save it." (Mark 8:35). John put it even more emphatically: "He who loves his life loses it; and he who hates his life in this world shall keep it to life eternal." (John 12:25). So when Arendt claims that Christianity reversed the ancient Greek evaluation of life and treated it as something sacred, she overplays one dimension of Christianity.

In fairness to Arendt, I should examine a particular feature of Christianity which appears at first glance to support her interpretation. This feature is the Christian belief that at the second coming of Christ to earth, the bodies of the deceased shall be resurrected, and for those who attain salvation, their bodies will live forever. This certainly gives the impression that Christianity does revere the body, but upon closer examination, it becomes apparent that the resurrected, immortal body of the saved is not the same body whose demands some Greeks found slavish. Augustine describes the resurrected body as follows:

> Just imagine how perfectly at peace and how strong will be the human spirit when there will be no passion to play the tyrant or conqueror, no temptation even to test the spirit's strength.... And what a body, too, we shall have, a body utterly subject to our spirit and one so kept alive by the spirit that there will be no need of any other food. For, it will be a spiritual body, no longer merely animal, one composed, indeed, of flesh but free from every corruption of the flesh.[38]

Given this image of the resurrected body, Christianity accomplishes less of a reversal of the Greek attitude toward life and its demands (as Arendt interprets that attitude) than a sanitization of life. If, indeed, the Greeks relegated biological necessity to the shadows of the private realm, Christianity promises to leave behind bodily demands and routines when the saved individual ascends into heaven. If Christianity holds life in high esteem, as Arendt claims, it is not the life of the mortal body that is esteemed, but the life of the spirit or soul.

In a later chapter this discussion of Christianity will be continued, although not in the context of a confrontation with Arendt. But the issues raised in the present confrontation—death and burdensome necessity—will also be central to that later discussion of Christianity.

To return to her interpretation of modernity, Arendt recognizes in the emergence of modern society a continuation of Christianity's reverent attitude toward physical life. She argues that the primary concern of society is the survival of the species and that the productive activity of humans today is organized by society in order to serve that end. As was the case with her interpretation of Christianity, there is some truth to her claim that the life processes are highly esteemed in modernity. I think Arendt is accurate when she points out that members of society are primarily concerned with "making a living," and that, at least ideally, society preserves the life of the species. But it is also the case here that Arendt's interpretation is somewhat one-sided and that she seems oblivious to certain attitudes and trends which run counter to her claim concerning the sacredness of life. What I have in mind here as countertrends are not the obvious ones, such as the proliferation of nuclear weapons or the nonmilitary degradation of the environment, both of which are social threats to the survival of the species which Arendt certainly recognized. Rather, I have in mind less obvious attitudes and trends, ones which are overlooked probably due to their mundane nature and to the fact that they occur in what remains of the private household.

Since Arendt seems to think that there is no private household in modernity (for Arendt, all that remains of privacy in modernity is the realm of intimacy, while the activities of the ancient household are now performed in society), it is not surprising that she would overlook these countertrends. However, vestiges of the ancient and medieval household have survived in modernity, and these traces have been infiltrated and organized by modern technology. An examination of what remains of the household is therefore in order, both because it will challenge Arendt's depiction of modernity by revealing a certain modern contempt for the life processes and because it will reveal something about the way in which technology has shaped the modern individual. In fact, the point I will be trying to make is that the fetishistic attitude toward technology which characterizes modernity is based on such contempt for the life processes. This should make obvious the tension between Arendt's interpretation of modernity and mine.

At the beginning of this chapter, I pointed out that in ancient Greece the household was the sphere in which production *and* consumption occurred. Even business activity was considered to be part of household production.[39] When Arendt claims that society has emerged as a sort of public household, she has in mind the productive dimension of household activity, and there is little doubt that this dimension has indeed become a social phenomenon. In the modern societies of the West and the East, productive activity is no longer organized by various private households, but by ever-larger corporate or state organizations. The productive life of the individual in both types of societies is spent not as a member of a household but as a worker or employee of an enterprise of much greater dimensions. When one emphasizes the consumptive rather than the productive activity of the household, however, it seems that the private household has not been completely lost to modernity, at least not in the West.

What remains of the household is its role as a center of economic activity. It is no longer the primary locus of economic activity as it was for Greek citizens, and modern economics, unlike the economics of ancient Greece, is no longer concerned primarily with the household. But the household, in an attenuated sense, remains an important site of economic activity. It is the site where consumption decisions are made and where financial resources are directed and distributed in order that the members of the household can consume what they need. Even more importantly, the modern household is the site at which lines of credit attach themselves to individuals and where obligations are incurred for the sake of consumption. While all this economic activity may be peripheral to the modern science of economics, marketing—the new science of the household—surely recognizes its importance.

In this capacity as the center of consumptive activity, the modern household takes various forms. The procreative dimension of the household has diminished in modernity, and motives other than the raising of children are often the impetus for forming household associations. Single-parent households have also become more common, and of course, the possibility remains for households of single individuals. What is central to the idea of a modern household is not that individuals of certain statuses have joined together, but that there exists some source of wealth or credit which is used to satisfy the consumption demands of the household. In this sense, a single person, a gay couple, and a group of individuals who share certain living expenses, are all examples of households. Heterosexual sex and children are no longer at the heart of the modern household, at least as I am using the term.

It is not only this consumptive activity that ties modern households to their counterparts in previous epochs; there is also present in modern households that element of necessity which played such an important role in defining the ancient household. And as was the case in the ancient household, necessity directs much of the consumptive activity of the modern variant. Of course, this is not to say that the content of necessity has remained the same over the centuries. While a certain set of needs—for food, clothing, and shelter—has remained necessary from ancient times to the present, necessity encompasses much more today than it did in the past. In fact, the content of necessity tends to continually expand in modernity, especially in capitalist societies, as what were once luxury items become necessities. For example, refrigerators and automobiles were once considered luxury items, but changes in the distribution of food, the neglect of public transportation, and shifts in the location of workplaces and housing have made such items necessary.[40]

Compared to the relatively limited needs of the premodern household, many modern needs hardly seem necessary, at least not in the narrow sense in which something necessary is indispensable for life. Life could certainly be sustained without automobiles or refrigerators. But one must beware of making distinctions such as real versus apparent needs, biological versus cultural needs, or even needs versus wants. Although there is undoubtedly some truth to such distinctions and in certain contexts they may be helpful (e.g., when one is trying to budget one's income), for the purposes of this argument such distinctions must be avoided because they conceal some important points.

The first of these points is made by Herbert Marcuse in a different context, where he stretches the meaning of "biological needs" to include culturally or socially generated needs. To quote Marcuse:

> I use the terms "biological" and "biology" not in the sense of the scientific disciplines, but in order to designate the process and the dimension in which inclinations, behavior patterns, and aspirations become vital needs which, if not satisfied,would cause dysfunction of the organism. Conversely, socially induced needs and aspirations may result in a more

> pleasurable organic behavior. If biological needs are defined as those which must be satisfied and for which no adequate substitute can be provided, certain cultural needs can "sink down" into the biology of man.[41]

Distinctions such as those listed in the preceding paragraph, therefore, obscure the fact that certain needs, despite their social origins, can become so deeply ingrained in the lives of individuals that they are as necessary or as real as any other need.

Another point, one which is crucial to this text, is also obscured by distinctions such as those just mentioned. Those distinctions, in making such a clear, unequivocal break between ancient needs and those which have emerged more recently, hide an important similarity that exists between ancient and modern necessity. This similarity is that modern necessity, despite its comparatively expansive nature, is grounded in the body—quite like ancient necessity. Of course I am not claiming that these types of necessity are identical in their relation to the body; they are significantly different. But to distinguish modern needs as artificial, unreal, or even unnecessary (in the narrow sense) misses the difference I have in mind, and it also conceals modern necessity's relation to the body. A more subtle, ambiguous, and undoubtedly debatable distinction is required here.

The distinction I would like to make between ancient and modern necessity is that ancient necessity was primarily concerned with satisfying the *demands* of the body, while modern necessity is largely focused on overcoming *limits* which are imposed by the body. By demands of the body, I have in mind needs such as those for food, clothing, and shelter—the needs which were identified by the rejected distinctions above as real, biological, or natural. And by limits of the body, I mean certain features of embodiment which are perceived as inconveniences, obstacles, or annoyances. Both these demands and limits will be discussed shortly, but first I must point out the difference between this demand/limit distinction and those others.

In identifying ancient needs as demands of the body, I am not trying to grant these particular needs a special, foundational status. The point in distinguishing the demands of the body from the limits which are imposed by it is not to set those demands apart as something irreducible or unavoidable, or to set them up as the measure of all other needs. In those distinctions I have rejected, however, such hierarchizing is usually the aim of the distinction. The point I am making is simply that the necessary activity of the ancient household revolved primarily around satisfying the demands which the body makes for food, clothing, shelter, water, and so on, and this is no longer the case in the modern household. A brief examination of the ancient Greek household will not only help to illustrate this point, but it will also indicate how Arendt's aversion to the body has affected her interpretation of classical Greece, the age against which she measures modernity.

In Xenophon's *Oeconomicus*, the activity of the Greek household is portrayed during the course of a Socratic dialogue concerning the principles of household organization and management. According to that portrayal, the activity which the wife supervised occurred within the shelter of the dwelling and was comprised of "the rearing of newborn children...the making of bread from the crop...the working of clothes from wool."[42] She did not have to perform these tasks herself, but it was her responsibility to oversee the labor of servants who performed them. It was also her responsibility to maintain order in the house by seeing to it that all tools and implements were returned to their proper place after being used[43] and that all provisions were stored properly and consumed at a rate which would ensure that they would not be prematurely depleted.[44] The wife was also supposed to look after the health of the slaves.[45]

It would seem, then, that Arendt was accurate in depicting the Greek housewife as one who was occupied primarily in the realm of the body. But according to the *Oeconomicus,* the husband was not as removed from the household and necessity as Arendt would have one think. Ischomachus, whom Socrates had questioned concerning the principles of economics, answered not only by recounting for Socrates the manner in which he educated and trained his wife for her role as supervisor of indoor activities, but he also discussed at much greater length his role as supervisor of the household activities which occurred outside the dwelling. Primary among these outdoor activities was farming, and this was true not only in the case of Ischomachus. Despite the developments which were made in various trades and crafts, classical Greece was predominantly an agricultural civilization. It has been estimated that during the fifth century B.C., nearly half of Athens' population lived in the countryside and worked the soil.[46] The husbands of these country households were responsible for producing at least enough olives, figs, and grapes to satisfy the demands of the household and, if possible, a surplus which could be sold.[47] Grains were also a staple of the Greek diet, but Athenian households were only able to produce one-quarter of the amount consumed.[48]

In ancient Greece, successful harvests depended on a great deal of attention and diligence on the part of the husband. The poor soil and arid climate of Greece required that fields be fortified and replenished frequently and that an extensive irrigation system be developed and maintained.[49] As with the indoor activity, most of these agricultural tasks were performed by slaves, but the successful farmer was an active overseer. Given Arendt's interpretation of classical Greece, one would expect these agricultural obligations to be resented by the husbands as an imposition on their freedom, but this is not the case. In fact, Socrates declares in the *Oeconomicus*, that "the pursuit of farming seems to be at the same time some soft pleasure, an increase of the household, and a training of the bodies so that they can do whatever befits a free man."[50] And it was not only the training of the body which made farming conducive to a good citizenry; because of their careful attention to the soil, Socrates expected husbands to be eager to defend the countryside against foreign aggression.[51]

It appears, then, that the attitude of the Greek citizen toward the household and necessity was not as harsh as Arendt portrayed it, and that the "gulf that the ancients had to cross daily to transcend the narrow realm of the household and 'rise' into the realm of politics"[52] was not so wide. The household activity of farming, at least, seems to bridge the gap between the household and the *polis*. On the one hand, farming was dictated by physical necessity, but on the other, it prepared citizens for political life. It improved and strengthened, and did not simply make possible, that political life. Other household activities of the husbands, which Socrates lumps together as "mechanical arts," were not so beneficial to political life.

Crafts and trades such as metal forging, potting, and cobbling were considered inferior forms of economic (or household) activity. Not only did such activity ruin the bodies of the participants, by requiring them to remain seated indoors for long periods of time or to work close to a fire, but it also deprived them of "leisure to join in the concerns of friends and of the city."[53] For these reasons, Socrates claims that such mechanics are "reputed to be bad friends as well as bad defenders of their fatherlands."[54] Nonetheless, citizens did participate in these mechanical arts, although not to the extent that *metics*, the resident aliens, did.[55] And when citizens did perform these arts, it was not simply as overseers; the owners of the various workshops of classical Greece often worked alongside the slaves and workmen.[56]

The development of such crafts and trades may appear to indicate that the business activity of ancient Greece had moved beyond the demands of the body, but this is not entirely true.[57] Although activities such as mining, gilding, instrument making (flute and lyre), and weapon making (sword and shield) were not directed by the demands of the body, many other mechanical arts were. The fuller and the cobbler were responding to the need for clothing; the potter provided utensils which made possible the transportation and storage of liquids such as olive oil, wine, and water; the carpenter and woodcutter provided shelter from the elements. Of course, the products of these craftsmen were not purely utilitarian. Athenian pottery, for example, was frequently graced by the black figures of the vase painters, and some pottery was never intended to store anything but rather was purely ornamental.[58] Nevertheless, a significant amount of the manufacturing activity of classical Greece was undertaken in response to the demands of the body.

This prevalence of bodily demands can even be recognized in the extensive trading activity of Athens. This trade was not primarily in manufactured goods, but was instead an agricultural exchange. The principal export of Athens was olive oil, a product of the husband, not the artisan.[59] During this period, olive oil was used not only as a food, but was also used as a fuel and a source of light.[60] The export of olive oil, and to a lesser extent wine and manufactured goods, was used primarily to acquire grain—usually corn—which was shipped back to Athens.[61] The maintenance of certain sources of imported grain and the control of the routes by which such grain made its way to Athens were a constant concern of the Athenians, and a crucial determinant of their imperial strategies.[62] In fact, it was in a battle to maintain the vital flow of grain from the Black Sea region that Athens lost the Peloponnesian War. In the battle of Aegospotami, the final battle of the war, the Athenians lost the vast majority of their fleet and control of the grain trade through the Hellespont. They were then quickly starved into surrender.[63]

The demands of the body, therefore, were of central importance in classical Greece and were not as shaded and hidden as Arendt makes it seem. The demand for food even played an important role in the public, political activity of Athens.[64] Indeed, it appears that classical Greece was not completely free of the "social housekeeping" which Arendt identified as a distinctively modern phenomenon. Since my primary concern here, however, is not to challenge Arendt's interpretation of classical Greek political life but to distinguish ancient and modern necessity, it is the necessary activity of the private sphere, the household, which must be emphasized. In the household of ancient Greece, the demands of the body held sway from the indoor routines of cooking, cleaning, and child rearing to the agricultural practices of the husband; they were even an important factor in the development of some trades and crafts. Therefore, a great deal of time in classical Greece was spent responding to the demands of the body, and this time was spent not only by women and slaves but by free men as well.

In contrast, members of the modern household spend much less of their time in the service of the body's demands.[65] Those demands are still satisfied by the household, but no longer through the time-consuming performance of certain reproductive tasks.[66] Rather, through the consumption of technological apparatuses or the products of such apparatuses, the modern household quickly satisfies the demands of the body.

Food, clothing, and shelter are no longer produced by the household, but are only consumed there. The production of these and other[67] necessities takes place in the "public/private hybrid" which Arendt called society. Although technology has certainly been crucial to the formation and development of this social production process, and that production process has in turn been an important influence on the modern self, the concern of this text lies not with the value which technology has for individuals as members of this production process but with the value which technology has for consumers.

In their productive activity, people often feel constrained in their relation to technology; they think of themselves as slaves to the machines;[68] in order to work, they have no choice but to use the newest technological developments. But in the consumption which occurs in the household, individuals tend to think of themselves as unconstrained consumers. In their consumption choices, they are no longer the slaves of technology but choose to use or not use technology freely. Because I think that this liberal notion of the sovereign consumer obscures one of the important ways in which technology shapes the modern self, I focus on the household consumption, rather than the production, of certain technological apparatuses. In the next chapter I will respond to those who would criticize this emphasis on consumption rather than production.

As an example of the way in which technology is consumed in the modern household, consider the case of a bodily demand which loomed so large for the Greeks—the demand for food. Unlike the Greeks, members of the modern household do not spend most of their time involved in the production, preservation, and preparation of food. Instead, the modern tendency is to buy food that is already prepared, refrigerate it at home until it is ready to be eaten, and then to use one of the various forms of instant heat to cook it. Not only is the food consumed in the modern household, but things like refrigerators and microwave ovens are also consumed in the satisfaction of the body's demand for food. And one can, of course, extend this list to include the various agricultural and transportation technologies which are indirectly consumed along with the food itself. In a later chapter I will examine the development of some household technologies, many of which are focused on the demands of the body, so for now let this example suffice. I merely want to point out here that the modern household consumes a wide array of technological products and apparatuses in order to quickly and easily satisfy the demands of the body.

It is no accident that I have used the issue of time to distinguish ancient and modern households, just as it is no accident that I have used terms such as *save, consume,* and *spend* to discuss the issue of time. Such economic terms are appropriate to a discussion of temporality because the issue of time is central to the economic activity of the household. In the ancient household, it was important that the way in which time was spent be properly organized and managed; it was through such careful attention to time that the satisfaction of the body's demands could be ensured.[69] The modern household, however, is less concerned with satisfying the demands of the body than it is with satisfying them quickly. The demands of the body are no longer thought of as requiring careful attention and proper planning. They are seen instead as inconveniences in that they limit or interfere with the use of time. The value of technology, I am arguing, lies in its ability to mitigate such inconvenience.

This modern attitude toward the demands of the body is *part* of what I was referring to earlier as limits imposed by the body. When such bodily demands are seen primarily as something impinging upon one's time, they become limits to overcome, rather than demands to satisfy. But as I indicated when I first introduced it, the demand/limit distinction is not unequivocal or unambiguous, and it does not neatly distinguish ancient and modern necessity. This is especially true in regard to the temporal limits of the body. As Arendt pointed out, the ancient Greeks were also concerned with saving time from the demands of the body. Instead of technological apparatuses, slavery and rigid sex roles were the means by which Greek citizens were able to free up some of their time for public activity. As I tried to show in my discussion of the Greek household, however, husbands also spent a great deal of time actively responding to the demands of the body, and this time was not considered wasted or ill-spent. Aristotle's contempt for the body and its demands does not appear to have been shared by the citizen farmers who constituted the largest part of the Athenian population. In fact, some scholars now argue that many Greeks found public, not private, activity to be distasteful or degrading.[70] In any case, the Greeks did not treat the demands of the body solely, or even primarily, as limits which had to be overcome.

The difference between ancient and modern necessity is not restricted to different attitudes toward the demands of the body, however. The limits of the body include more than just those temporal limits which are imposed by the demands of the body. Rather, modern necessity finds in the body an array of limits, some of which are not so much temporal as spatial. As embodied beings, humans exist in a world in which other things, including other persons, are dispersed in a spatial field. Put differently, the body of the human delimits a given space, and other things—things other than the self—are at a distance. Overcoming this distance, by moving either persons or things, is a concern for people as embodied beings.

This need to move people and things, which I will call the need for conveyance, is certainly not a uniquely modern need. The Greeks, after all, were excellent sailors and were able to establish an empire and import grains over great distances. And before them, the Egyptians had moved large blocks of stone to the sites of the pyramids and then put them into place.[71] But while these ancient civilizations were indeed concerned with the need for conveyance, this need held a subsidiary or derivative status for them. By this I mean that the need for conveyance was important for ancient civilizations inasmuch as distance was a hindrance to the satisfaction of other needs. The Greeks' sailing prowess, for example, was largely a response to their need for grain. And in the case of the Egyptian pyramids, the need to erect an immortal monument to the Pharaohs was the impetus for the marvelous movements accomplished by that civilization.

In modernity, however, distance is no longer treated as an environmental feature of embodiment; rather, distance is another limit which is imposed upon people by their bodies. The need for conveyance—the need to overcome the spatial limit of distance—is a primary need in modernity. No longer is conveyance merely a question of the ability to move *what needs* to be moved to *where it needs* to be; movement today is necessary in and of itself, and any impediment to movement is an obstacle to be overcome or assaulted by technology.[72] Air travel is an obvious example of the use of technology to overcome one of the chief impediments to movement—gravity. Another example is telecommunication, which allows the conveyance of information over great distances almost instantaneously. Other examples of modern technological conveyance will be examined in a later chapter.

This attitude toward distance as a spatial limit which is imposed by the body is exacerbated by the modern attitude toward temporal limits. A whole group of needs has emerged around the point where the concern for saving time merges with the disdain for the limit of distance. As soon as a spatial barrier has been overcome, a new set of temporal limits emerges around this achievement. Again take as an example the ability to overcome gravity and fly from place to place. Once this breakthrough was attained, it became necessary not only to fly wherever people wanted (i.e., to overcome all spatial barriers to flight), but it also became necessary to fly as frequently and as fast as they needed. Time spent traveling is considered an inconvenience and must be constantly lessened by technological developments.

The need for speed, both in conveyance and in people's ability to satisfy the demands of the body, is a hallmark of modern necessity.[73] The need for speed also helps to explain the continuously expanding range of modern necessity. Unlike purely spatial limits, as soon as a speed limit is overcome, another limit is simultaneously established. The need to do things and get places as quickly as possible is a need that can never be satisfied. Every advance imposes a new obstacle and creates the need for a more refined or a new form of technology.

The point of this discussion of modern necessity is to reveal that, despite its expansive and apparently nonbiological nature, modern necessity, like ancient necessity, is based upon the body. However, the modern attitude toward the body, as it is reflected in the consumptive activity of the household, is quite different from the ancient Greek attitude toward the body. While the Greeks thought that the satisfaction of bodily demands required careful attention and planning throughout the household, modernity treats the body instead as the source of limits and barriers imposed upon persons. What these limits require is not planning and attention, but the consumption of various technological devices that allow people to avoid or overcome such limits.

The value of technology for the modern household, therefore, lies in technology's ability to mitigate the effect of bodily limits. The word I choose to express this value is *convenience.* The appropriateness of this choice is indicated, in part, by the simple fact that the various technological apparatuses which are consumed by the household are often called 'modern conveniences.' Items such as automobiles, dishwashers, and telephones are conveniences in the sense that they make life easier or more comfortable. A more important indication of the appropriateness of the word *convenience,* however, is that this sense of the word—in which convenience means ease and comfort—is a uniquely modern sense.

The noun *convenience* and the adjective *convenient* are Latin in origin.[74] *Convenience* is an adaptation of *convenientia,* which means 'meeting together, agreement, accord, harmony, conformity, suitableness, fitness.' The adjective *convenient* is based on the present participle of the verb *convenire,* which means 'to come together, meet, unite, agree, fit suit.' Prior to the seventeenth century, the meanings of the English words remained quite close to these Latin roots. Something could be described as convenient or as a convenience if it was in accordance or agreement with something such as nature or 'the facts,' or if it was suitable or appropriate to a given situation or circumstance, or if it was morally appropriate. These pre-seventeenth-century meanings, however, are now considered obsolete.

The modern meaning of *convenience* is 'the quality of being personally convenient; ease or absence of trouble in use or action; material advantage or absence of disadvantage; commodity, personal comfort; saving of trouble.' And the current sense of *convenient* is 'personally suitable or well-adapted to one's easy action or performance of functions; favourable to one's comfort, easy condition, or the saving of trouble; commodious.'

The difference between the obsolete and current meanings of these words lies not only in the modern addition of the sense of ease and comfort, but also in the fact that what remains of the older meaning's sense of suitability has shifted and narrowed. Convenience is no longer a matter of the suitability of something to the facts, nature, or a moral code; suitability in the modern meaning of convenience refers back to the person, the self. Something is a convenience or convenient in the modern sense of these words if it is suitable to personal comfort or ease. This shift in the reference of convenience corresponds to the change in attitude toward the body which occurred around the same time. The attitude toward the body as the source of burdensome limits is reflected, however obliquely, by the modern meaning of convenience. After the seventeenth century, something is a convenience if it is suitable to the modern task of overcoming the limits which are imposed by the body.

Another etymological shift must also be noted here. The word *comfort*, which is central to the modern meaning of convenience, underwent a corresponding change in meaning, although it appears that this occurred perhaps as early as the fifteenth century.[75] Prior to that point, the principal meanings of *comfort*, in either its verb or substantive form, were centered upon strength and support. To comfort, or be a comfort, meant to support, strengthen, or bolster, in either a physical or mental sense. During this period, there also was a sense of comfort as the removal or absence of pain or discomfort, but this sense was limited to mental distress. It was not until the fifteenth century that the verb comfort included the sense of removing physical pain or discomfort. And the substantive sense of comfort as 'a state of physical and material well-being, with freedom from pain and trouble, and satisfaction of bodily needs,' was not widespread until the nineteenth century. It is in this later, bodily sense that the word *comfort* is used in the definition of convenience.

These etymological shifts, of course, are hardly conclusive proof of any change in attitude toward the body; indeed, it is doubtful that such a change could ever be conclusively proven. But the changes in the meanings of *convenience* and *comfort* are valuable as linguistic traces of that other change. In the following chapters I will examine other inconclusive forms of evidence, such as developments of certain technological apparatuses of the modern household, changes in religious ideas and doctrines, and innovations in political thought.

Before moving on to these other areas, however, a final thought on Hannah Arendt must be offered. It may appear that I have gone out of my way to challenge not only her interpretation of modernity but her interpretation of classical Greece and Christianity as well. The point of challenging Arendt's various interpretations, however, has been to reveal that despite her reputation as an unswerving critic of modernity, she did harbor a particularly modern trait or tendency. This trait, of course, is that attitude toward the body which I have cited as a distinctive feature of modernity. While Arendt did not celebrate the technological progress of modernity and did not appear concerned with the various limits of the body, she did tend to treat the body as a hindrance or inconvenience to public life. Throughout *The Human Condition*, Arendt assiduously avoids, if she does not exactly overcome, the body. In her interpretation of ancient Greece, Christianity, and modernity, Arendt displays a certain unwillingness to spend or waste time examining the private, bodily realm.

While Arendt would have argued that her attitude toward the body was influenced by the Greeks and was therefore diametrically opposed to anything modern, I tried to show in my discussion of ancient necessity that the public and private were not as distinct as Arendt liked to believe, and that the demands of the body played a significant role in both the public and private realms of ancient Greece. Arendt's aversion to the body, I am suggesting, kept her focused on Aristotle's derogatory claims about the body, necessity, and privacy, and kept her from looking more closely at the necessary activity of Greece.

This same attitude toward the body prevented Arendt from noticing the essential ambivalence of Christianity's attitude toward the body and the life processes. By making the suffering, toil, and eventual death of mortals essential to the attainment of an otherworldly immortality, Christianity sanctified mortal life, according to Arendt. And it was this disaster which Arendt found to be the most consequential accomplishment of Christianity. But this interpretation completely overlooks the connection between guilt and the body in Christianity and the fact that mortal life was considered a punishment for sin. For Arendt, however, the mere association of the life of the body with immortality could be nothing other than the disastrous sanctification of life.

Of course, it is the influence of this attitude toward the body on Arendt's interpretation of modernity that is most important for my argument. Her aversion to the body causes her to focus on the quasi-public, or social, production of necessities in modernity. But even here she is so repulsed by the public display of base necessity that she interprets it as modernity's reverence for life and never closely examines what necessities are actually produced in modernity. And that same attitude prevents Arendt from looking more closely into what remains of the private household. A more detailed examination of the household and modern necessity, however repulsive, might have revealed to Arendt the extent to which modernity is closer to her Greek ideal than was Greece itself. She might have realized that, far from sanctifying the body and the life processes, modernity is distinguished by the ability of the masses to free themselves from the limits of the body through the ravenous consumption of technology. The trajectory of modernity is to render everyone free not only from the limits which are imposed by the body, but even from the body itself. (I will get to this point much later.)

The interpretation of *The Human Condition* I have offered here is, of course, highly ironic. For Arendt and I begin from similar concerns. In the Prologue to *The Human Condition*, Arendt discusses the space-age attitude toward the earth as "a prison for men's bodies" and the attempt by scientists to create life in a test tube. I share this concern about the direction, or trajectory, of modern technology. And while Arendt claims that no answer to these "preoccupations and perplexities" is offered in her text, she does suspect that these phenomena are grounded in a desire to escape the human condition.[76] I agree with Arendt on this much, but from my perspective, "the very quintessence of the human condition" is not the earth, as Arendt claims, but the body. The irony, therefore, is that Arendt's treatment of the body as something which should properly be hidden in private appears to me as a version of the modern attempt to escape the human condition.

CHAPTER 3

Marxist Perspectives on Consumption

In the last chapter, I mentioned in passing that Arendt recognized Marx as an unwitting smuggler of Christian ideas. By this remark, I meant that Arendt saw in Marx's concern for the social dimension of the production process and his appreciation for the development of these social forces a modern form of Christianity's reverence for life. I, too, am troubled by this Marxist emphasis on production, but for very different reasons than Arendt. What concerns me is not that this focus on production deflects attention away from the loss in modernity of any public sphere in which action might occur (although it does do that); rather, the Marxist preoccupation with the capitalist production process disturbs me because it diverts critical attention away from the consumptive activity of the modern household. In this sense, I see an effective similarity between Arendt and the Marxists from whom she sought to distance herself: both obscure an important way in which technology shapes the modern self. And while I am aware that this criticism would bear more directly on Arendt than on Marxists—since she is concerned more with the effects of technological development than with economic exploitation—I nonetheless think that my point has some bearing on current Marxist thought.

Even if the primary task of Marxist thought is to uncover and eliminate economic injustice and exploitation, there is no doubt that the tenacity of capitalism in the late twentieth century is bound up with the technological fetishism of modernity. It is no longer enough to point out that the capitalistic production and exchange of things as commodities conceals the productive relations among men; that is, thought which would challenge capitalism today can no longer remain satisfied with Marx's revelation of the 'secret' of commodity fetishism.[1] In the latest stage of capitalism, the character of the commodities themselves must be closely examined along with the needs which the consumption of those commodities satisfies. For the technological character of the commodities consumed in late capitalism harbors a secret of its own, a secret which may help explain not only the dominance of technology but also the resiliency of advanced capitalism.

In making this claim that Marxist interpretations of modernity overemphasize the production process and neglect the realm of consumption, I am aware that significant steps in the opposite direction were taken by several Marxist theorists in the 1970s. In fact, it is those theorists in particular whom I have in mind when I make this claim.[2] Before I examine some of those steps away from the rigidity of structuralist Marxism, I should point out that the claim I am making here applies less to Marx than it does to Marxists. Writing in England in the nineteenth century, a period in which various facets of the production process were being mechanized, Marx's focus on the nature of the production process is hardly exceptionable. And even so, Marx's appreciation for the transformation of consumption which was required by capitalist relations of production is a significant, if neglected, element of his thought. After my examination of some recent Marxist thought on consumption, I will examine in more detail Marx's treatment of the transformation of consumption and try to bridge the gap between Marx and that recent thought.

The first of the three Marxist texts I will examine is *Capitalism, Consumption and Needs*, a collection of four essays by Edmond Preteceille and Jean-Pierre Terrail. Three of the essays were originally published in French in 1977 and the fourth in 1985. In the first essay, "Commodity Fetishism and the Ideal of Needs," Terrail indicates what may be considered a very plausible explanation for the Marxist aversion to any protracted examination of needs and consumption.

> The concept of need is inherent in the vulgar realism of bourgeois ideology,...it is always...taken to be the very essence of human existence...; indeed, the whole structure of economic liberalism is the work out *from this premise*. The free expression of the needs of the free worker in the market-place, in the sphere of consumption—this bourgeois vision of the highest degree of freedom depends on a close link between needs and consumption.[3]

Despite this bourgeois, or liberal, aura which surrounds the issue of needs and consumption, and the risk one runs of reinforcing liberal ideology by merely raising this issue, Preteceille and Terrail insist that "the question of needs is at the heart of social conflict."[4] Making the same point I have already made concerning Marx, they recognize that in the nineteenth century Marx may have had a point in claiming that the question of needs was out of date, "but growth, opening the way to abundance, had swept all that aside."[5] Therefore, as the title of their text suggests, Preteceille and Terrail face squarely up to this issue.

To avoid the dangers which attend any discussion of needs and consumption, Preteceille and Terrail employ two safeguards that keep them from implicitly endorsing any version of the sovereign consumer, which for liberalism is the source of all needs. The first of these safeguards is a recognition of the historicity of the consuming subject and its needs.[6] The liberal notion of the sovereign consumer completely neglects the historical specificity of that consumer and does not recognize that the idea of the free consumer in the market place is something which emerged in the course of history.[7] Furthermore, the neglect of the historicity of consumption and needs obscures the way in which the needs of the consumer continue to be influenced by historical developments. By paying close attention to the historical forces shaping the modern consumer, Preteceille and Terrail lessen the risk of falling back into the liberal idealism of the consuming subject.

But a concern for history by itself is not enough to prevent a lapse into vulgarity. Preteceille and Terrail point out that since the end of the nineteenth century, liberal thought has tried to come to grips with its ahistorical tendency by recognizing the influence social developments have on needs and consumption.[8] Although the idea of a 'consumer society,' the upshot of this rethinking of consumption, does take into account the historicity of needs, it still remains objectionable to Preteceille and Terrail. This is because it retains another element of liberal thought—the relegation of production to a merely instrumental role vis-à-vis consumption.

From the liberal perspective, the production process exists only as an instrument that satisfies the needs of the realm of consumption. It makes little difference to Preteceille and Terrail whether that consumptive realm is perceived as isolated individuals or as a consumer society. In either case, "the logic of consumption appears as the primary, determining, autonomous moment, while productive labour is reduced to a simple instrument for provisioning the market."[9]

To avoid falling into this flawed logic, Preteceille and Terrail rely on a second safeguard: the rejection of any "autonomisation of the spheres of consumption and need" in relation to the sphere of production and the reversal of "the order determination between production and consumption established by vulgar economics."[10] It is not an autonomous sphere of consumption and needs that determines the production process, but the production process that ultimately determines consumption and needs. For Preteceille and Terrail, the determination of needs "can only be understood in one way: the needs that production satisfies are the needs of production itself, the demands of its reproduction."[11] And the production process not only produces needs, but consumers as well: "well-defined social agents...historical forms of individuality made up of a whole body of inclinations and capacities...."[12]

There is, of course, nothing new here. Taken together, these two safeguards amount to no more than the historical materialism Marx outlined in *Grundrisse*.[13] So instead of distinguishing these safeguards, I might have simply written that their historical materialist approach to the issue of needs and consumption keeps Preteceille and Terrail from sliding into what they call a "substantialism" of the needs of consumers. However, the point in breaking up these elements of historical materialism (which I am sure is a capital offense in some circles) is to pose the question whether both of these elements are necessary to prevent an examination of needs and consumption, and the values which influence them, from falling into liberal idealism. It seems to me that an awareness of the historicity of the consumer and its needs is essential to this preventive purpose. But it also seems that the insistence that production is always the dominant historical force, that consumption and needs are in the last instance determined by the mode of production, has become a hindrance to understanding the development of needs under the advanced capitalism of modernity. Perhaps, like the autonomous individual of liberal economic thought, this aspect of Marx's thought is a historically specific idea which has come to obscure some recent developments. (Could it be anything but a historically specific idea?) Preteceille and Terrail seem to approach this point of view when they claim that:

> A proper insistence on the determining character of the social relations of production has overshadowed not only the necessary analysis of the specific structure of modes of consumption, but also an analysis of the relations between the two spheres, which has been reduced to a single, mechanistic determination.[14]

What is needed, claim Preteceille and Terrail, is a theoretical initiative to move beyond such analysis, but in the end, their attempt to move beyond an oversimplified view of the relation between production and consumption remains anchored to Marx's idea of the ultimate dominance of production.[15]

Before briefly examining the progress Preteceille and Terrail have made in their theoretical initiative, I should emphasize that when I suggest that it might be time to move beyond this particular element of Marxist doctrine, I am certainly not advocating a reversal of the Marxist position, which would restore consumption and needs to the determinant position they held in liberal economics. In each of the preceding chapters I indicated that the "sovereign consumer" is a major obstacle to understanding the hold which technology has on modernity. Nor am I arguing that material conditions are no longer important for an investigation of modern needs. In fact, I am eventually going to claim that the Marxist preoccupation with production has caused some thinkers to ignore other important material considerations. The claim I am making here is simply that an analysis of the most recent developments of needs must consider influences other than—or rather, alongside of—the capitalist production process. The foregone conclusion that needs are always a function of the production process seems to me to be as misleading as the liberal alternative that it is supposed to counter.

To get back to Preteceille and Terrail's analysis of needs, their progress beyond an oversimplified view of production/consumption relations lies in their recognition that a given mode of production can generate needs which in turn effect, even challenge, that mode of production. This uneven, or skewed, reciprocity between production and consumption is possible because, even though the mode of production is the dominant force in any complex social structure—or "structure in dominance," to borrow Althusser's phrase[16]—the mode of production itself is not monolithic. Rather, there are two distinct elements in a mode of production: the forces of production, which correspond to the productive capacity of a given society, and the relations of production, which comprise the organization of that society.[17] Social needs emerge when the reproduction of both the forces and the relations of production becomes problematic, or contradictory. Preteceille and Terrail go so far as to claim that it is "impossible to consider the historicity of needs without referring them to the logic of the mode of production as a contradictory union between relations of production and productive forces."[18]

Preteceille and Terrail use the English Factory Acts as an example to illustrate this point.[19] In the middle third of the nineteenth century, productive forces were greatly increased by the mechanization of various productive activities. Under the existing relations of production, however, in which laborers had no choice but to work for the wages, during the hours, and under the conditions determined by each individual employer, the increased productivity supplied by mechanization soon came to threaten the reproduction of the mode of production itself. In order for the owners of the mechanized operations to recoup the value expended in acquiring this machinery, they had to extend the working day to the limit of human capability. It was imperative that the capitalists recover this value before mechanical innovations made their newly purchased equipment obsolete or comparatively inefficient.[20] By increasing the number of hours worked in a day, the owners were able to increase their daily share of surplus-value and thereby quickly recover the value of their increased outlay of fixed capital.

Under this mode of production, the labor force was eventually exhausted as workers were pushed to their limit and then replaced when used up. The reproduction of the forces of production, however, requires a supply of labor power which is able to reproduce itself over a long length of time. This contradiction between the forces of production, which were enhanced by the introduction of machinery, and the relations of production, which allowed the unlimited extraction of surplus-value in order to pay for this machinery, threatened the mode of production itself, and the need arose for some limitations on the length of the working day. So although the demand for such legislation was first articulated by the laboring class in opposition to the class of owners, it seems clear that the need to be satisfied by the Factory Acts was ultimately a need of the mode of production. That groups of capitalists ultimately began to call for such restrictions would seem to bear this out.[21]

The legislative regulation of the relations of production, however, did not simply make possible the reproduction of the forces of production. Rather, passage and enforcement of the Factory Acts "contributed to the rapid introduction of mechanization, and gave impetus everywhere to the acceleration of technical developments and the intensification of labour."[22] By restricting the number of hours that could be worked in a day, the Factory Acts also restricted the amount of surplus-value that owners could appropriate daily. This restriction on the accumulation of surplus-value made it necessary to increase the productivity of the forces of production even further. More powerful engines, faster machines, and a more disciplined, routinized work force would allow the owners to increase the amount of value produced per hour, thereby offsetting the stifling effect the Factory Acts had on the accumulation of profits.[23] In turn, this intensification of the labor process produces anew the need for a shortening of the workday, which requires further advances in the forces of production, and so on.

The interplay between the mode of production and the needs it generates is well illustrated by Preteceille and Terrail's example of the Factory Acts; the need for some limitations on the length of the workday brought about changes in the forces of production. Preteceille and Terrail find in this interplay between needs, or consumption, and production a source of hope for the eventual dissolution of the capitalist mode of production. The needs generated by the contradiction between the forces and relations of production drive the mode of production to higher stages of development, and the needs which have emerged thus far in the latter half of the twentieth century may require the elimination of capitalistic relations of production themselves.[24] They cite as examples the recently articulated needs for

> a slower pace of work...better living conditions...diminution of travelling time and an increase of comfort...an improvement in the health system...antipollution measures...the application of technical progress to the benefit of living labour...[and ultimately] some sort of social control by the workers themselves.[25]

Needs such as those listed above pose a challenge to capitalistic relations of production because they cannot be readily satisfied through individual, private ownership and appropriation of commodities. Rather, these needs require for their satisfaction a "socialization of consumption." By this, Preteceille and Terrail do not mean the absolute elimination of private ownership and appropriation, but the transcendence of this form of consumption "at the point where it becomes an obstacle to reproduction and the development of productive forces."[26] Where the working class is unable to attain what it needs to reproduce its labor power—whether these needs are for adequate housing or increased leisure time—collective consumption facilities are emerging to satisfy these needs. Since Preteceille and Terrail focus on this sort of need and recognize the tendency toward socialized consumption, it is not surprising that they call not only for the satisfaction of the existing needs of the working class, but also "for the expansion, development and transformation of those needs themselves. Breaking up capitalist hegemony entails an explosion of needs...."[27]

To an American in the early 1990s, this tendency towards socialized consumption may appear outdated, and Preteceille and Terrail's confidence in the revolutionary potential of the proliferation of needs may seem unfounded. But Preteceille and Terrail are not oblivious to the possibility that some of the social needs they identified might be satisfied through individual, private consumption or that increases in the private consumption of commodities might deflect attention from those social needs altogether.[28] Nor are they unaware that there is a countertendency to the socialization of consumption which strives to 'privatize' those collective consumption facilities which have already been established.[29] On the contrary, they recognize that through

> the private character of commodity consumption...capital imposes practices (and the values implicit in them) which reinforce its ideological and practical dominance; the objects of consumption can be seen as representing so many ideological messages, which have underlying them as many constraints leading to competitive individualism, to the depoliticisation, fragmentation and opposition of the dominated classes.[30]

Despite this recognition of the counterrevolutionary potential of private consumption, Preteceille and Terrail never closely examine the "ideological messages" or implicit values conveyed by objects and practices of consumption. They point out the danger that such private consumption presents to their socialist objectives and then quickly return to the promise of social needs. This can be explained in part by the fact that Preteceille and Terrail take France, not the United States, as their model of a capitalistic society. As they point out in some statistical detail, France in the late 1970s was far removed from the 'myth' of American consumer society.[31] In France, the threat from private consumption may be weaker than in the United States. Another element of an explanation can be found in Preteceille and Terrail's dialectical perspective. For them, there is an inevitability to the development of social needs, an inevitability grounded in the contradictions of the capitalist mode of production and their dialectical resolution. "In the long run," claim Preteceille and Terrail, "consumption practices cannot avoid, and will in fact be less and less able to avoid, the class confrontations which owe their meaning and bearing to the logic of relations of production."[32] Just as 'free market' relations of production had to give way in the face of the working-class need for state regulation of the labor process, so too will the exploitative relations of late capitalistic production, along with its individualistic consumption, have to give way to the social needs produced by that mode of consumption. There is no stopping the force of the dialectic.

In any case, there is little doubt that private consumption practices and the objects of consumption can be interpreted along the lines of Preteceille and Terrail's argument, even if they have not done so. Michel Aglietta, however, in *A Theory of Capitalist Regulation*, has done just that. Although Aglietta's text does not focus primarily on developments in consumption and needs, but instead offers a much broader view of late capitalism than Preteceille and Terrail's *Capitalism, Consumption and Needs*, it nonetheless makes important strides toward understanding the way in which technology has infiltrated the modern household.

I should mention at the outset of this discussion of Aglietta another important difference between his work and that of Preteceille and Terrail. As indicated by the subtitle of his text—*The US Experience*—Aglietta takes the United States, not France, as his model of late capitalistic society. He explains this choice in his introduction: "The particular selection of the United States is designed to highlight the general tendencies of capitalism in the 20th century. The USA, in effect, experienced a capitalist revolution from the Civil War onwards."[33] The outcome of this revolution, claims Aglietta, was the establishment of "the most adequate structural forms for perpetuating capitalist relations of production that the class struggle has yet created anywhere."[34]

One of the main features of this capitalistic revolution was the development of a "social norm of consumption," but what Aglietta is referring to by this norm is not the potentially revolutionary social needs which Preteceille and Terrail identified. On the contrary, this social norm of consumption is one "in which individual ownership of commodities governed the concrete practices of consumption."[35] In his discussion of consumption, Aglietta focuses on the objects or commodities consumed by individuals in the modern household. And coming even closer to the perspective I am developing here, he also stresses that the evolution of the social norm of consumption of the United States "was governed by the replacement of direct activity at home by time-saving equipment,"[36] which I would call conveniences.

According to Aglietta's Marxist perspective, of course, the development of this norm of consumption is ultimately an effect of the production process. It constitutes one part of what he, borrowing and developing Gramsci's term, calls "Fordism." This is the term Aglietta and Gramsci use to describe the "semiautonomisation" of the labor process which occurred in early twentieth-century America. This development of the production process can be explained in the terms of the previous example of the English Factory Acts.

Through that example, it was shown that the mechanization of the labor process led to the workers' need for a shortening of the workday, and how the satisfaction of this need led to further intensification and mechanization of the workplace. Although the eight-hour day was not established by the U.S. government until 1938, various American labor unions had been struggling since the 1840s, with uneven success, to shorten the workday.[37] And alongside these efforts to alleviate the strain of the mechanized production process, American workers were also able to restrict the output of that process from within. Because American labor unions of the nineteenth century were generally craft unions, the workers were able to retain their knowledge of the various labor processes, and use this knowledge to exert some resistance to capital's attempts to speed up the pace of production.[38] This resistance, of course, limited the owners' ability to quickly recover the value of the machinery in which they had invested and, as in England, owners sought to increase the rate at which they accumulated profits by further intensifying and mechanizing the production process.

The first American attempt to intensify the production process, which occurred around the turn of the century, goes by the name of "scientific management," or Taylorism, after Frederic Winslow Taylor, forerunner in the field. Briefly, Taylorism sought to rationalize the labor process by gathering from workers all knowledge and information concerning that process and making such knowledge the exclusive domain of managers. These managers could then reorganize the production process according to that knowledge with the intention of eliminating all waste of time and motion.[39] Fordism, named after Henry Ford, further developed these intensification techniques of Taylorism, and combined them with innovations in mechanization—i.e., the continuous assembly line. With the semiautomatic assembly line, management was able to control and synchronize the entire production process. As Aglietta puts it, "The individual worker thus lost all control over his work rhythm.... In this mode of organization workers are unable to put up any individual resistance to the imposition of the output norm, since job autonomy has been totally abolished."[40]

In his essay, "Americanism and Fordism," Antonio Gramsci pointed out that Fordism not only "rationalized" the productive activity of the assembly-line workers, but it also sought to control their activity outside the workplace. The regularity of this new phase of production required that workers' performances be consistent all day long, from day to day. The industrialists' attempt to prohibit the consumption of alcohol and their exhortations against sexual licentiousness served "the purpose of preserving, outside of work, a certain psycho-physical equilibrium which prevents the physiological collapse of the worker, exhausted by the new method of production."[41] The prohibitions of both the consumption of alcohol by the workers and the consumption of the workers by excessive sexual activity were required by the assembly line.

While Gramsci emphasized this prohibitory element of Fordism, Aglietta points out another face of Fordism's relation to consumption. The demands of the assembly line required not only that certain consumptive activity be prohibited, but also that other forms of consumption take place. By eliminating any lulls or gaps in the working day, Fordism made it necessary that all recuperation or rejuvenation of the work force take place outside of work. To quote Aglietta:

> The increased exhaustion of labour-power in the labour process had to be entirely repaired outside the workplace, respecting the new time constraint of a strict separation between working and non-working hours.... Individual commodity consumption is the form of consumption that permits the most effective recuperation from physical and nervous fatigue in a compact space of time within the day, and at a single place, the home.[42]

The needs of modern workers for various time- and labor-saving commodities—the social norm of consumption—can convincingly be interpreted as needs of the production process. High-speed assembly-line production requires such consumption in order to reproduce a stable labor force. But there is also another sense in which this consumption norm is needed by the production process. Aglietta explains that the continuous consumption of commodities by the workers enables capital to overcome the disjuncture which had often occurred between that section of the production process which produced the means of production, such as machinery (Department I), and that section which produced the means of consumption, such as household appliances (Department II).

Prior to the emergence of Fordism, developments in Department I were sporadic and uneven, and each new development in the means of production was characterized by a massive increase in fixed capital expenditures which was followed by a depression in such capital formation. The reason for this depression in capital formation was that the exchange of consumer items in Department II did not keep pace with the productive activity in Department I, and therefore the demand for the new means of production by Department II was not great enough to permit the firms in Department I to recover the value of their fixed-capital investments.[43] Of course, the outcome of this disharmony between Departments I and II goes beyond the rate of capital formation. In periods of extreme economic disjunction, the "means of production are destroyed on a massive scale right across society."[44] This whole cycle begins again with the development of new techniques of production.[45]

With the emergence of the social norm of consumption under Fordism, however, this cyclical pattern of capital formation was to a great extent eliminated. The steady consumption of soon-to-be improved commodities made it possible for the means of production to be depreciated and eventually replaced in a gradual, stable manner. As Aglietta describes this important effect of the social norm of consumption:

> The fundamental fact is that the qualitative transformation of the forces of production has become a permanent process, instead of being chiefly condensed into one specific phase of the cycle of accumulation. This change is due to the interaction of the two departments of production; each now provides the other with its markets as they combine to lower the value and diversify the commodities of mass consumption. Obsolescence becomes generalized and permanent.[46]

The significance of Fordism, therefore, lies in this integration of consumption and production. The continuous consumption of various conveniences by the work force provides capital not only with a stable, well-rested supply of labor power, but it also allows the two main sectors of the economy to synchronize their productive activity. This situation is what Aglietta referred to as "the most adequate structural form for perpetuating capitalist relations of production." But Fordism, which came into full bloom after World War II, eventually ran up against some limits in the late 1960s, and the nature of these limits brings Aglietta's analysis close to that of Preteceille and Terrail.

The private, individualistic consumption which flourished in the United States under Fordism required a corresponding expansion of the role of the state as the guarantor of the continuity of consumption. It was necessary for the state to ensure that, in periods of economic dislocation, individuals could continue to consume and could still meet financial obligations already incurred through previous consumption. Aglietta points out how "this implied legislative arrangements, a homogenization and socialization of wages, and the establishment of social insurance funds against the temporary loss of direct wages."[47] A pension system for retired workers was also required in order to maintain the consumptive activity of this significant segment of the population.

While the satisfaction of these 'social needs' was provided to a great extent by the state (in the form of the New Deal), some of these needs were met wholly or in part by the development of private pension funds and insurance plans. Private pension funds, which were usually developed through labor's collective bargaining with management, supplemented the public system of Social Security. Private medical insurance has been the primary response in the United States to the need not only to pay medical expenses, but also to provide income during recuperation. Aglietta points out how these private responses to social needs strengthen the position of capital, since the enormous amount of value that is accumulated in these private plans is administered by capital itself.[48] While this point is important for understanding the strength of capitalism in America, as well as the current pressure to 'privatize' public responses to social needs, the existence of such private responses has not allowed Fordism to avert its crisis.

Whether the social needs which underlie the Fordist norm of consumption are satisfied publicly or privately, the cost of such 'socialized consumption' is ultimately paid out of the surplus value which is available to capitalists. If these programs and services are provided by employers as part of a collective bargaining agreement, the cost amounts to indirect wages paid to the workers. If these services are provided by the state, the cost either causes an inflation of wages, which are then taxed more heavily, or profits are taxed in a more direct manner. "In either case," claims Aglietta, "there is a restriction on relative surplus-value and consequently an obstacle to the law of accumulation."[49]

Furthermore, the cost of these various preconditions of the social norm of consumption tends to increase as the semiautomatic production process progresses. This is because the mechanized production process is unsuitable to the provision of these collective goods and services. The savings that capital is able to extract from labor costs by using the assembly line are unavailable in the area of collective services, and the provision of these services becomes comparatively expensive as costs in commodity-production decline.[50]

The increasing costs of collective consumption are not a problem as long as capital continues to increase the rate at which it is able to extract profits from the mechanized production process. But eventually, Aglietta claims, that rate of accumulation or profit reaches its limit, as workers begin to resist management's attempts to increase productivity through the further fragmentation and mechanization of the production process. Aglietta identifies the mid-1960s as the point at which labor's resistance began to halt the decline in real wage costs that had been achieved by Fordism.[51] At this point, the costs of social consumption were no longer offset by increasing profits, and those costs became an unbearable burden for capital. From Aglietta's perspective, therefore, "it is not surprising...that the crisis of Fordist work organization should at the same time have been the occasion for a general drive of the capitalist class to curtail social expenditures, and have ushered in a period of retrenchment in public finances."[52]

Ultimately, Aglietta's analysis leads him to a position that appears close to that of Preteceille and Terrail. His conclusion that under the Fordist production process "[t]he *socialization of consumption* becomes a decisive terrain and battle-ground of the class struggle"[53] echoes Preteceille and Terrail's claim that "the question of needs is at the heart of social conflict."[54] But there is an important difference. For Aglietta, who focuses on the United States, socialized consumption and social needs are the preconditions of a private, individualized norm of consumption, whereas for Preteceille and Terrail, who focus on France, such needs are a radical alternative to private consumption. This is not to say that Aglietta does not recognize the potential challenge social needs can pose to capitalism; it is simply that the realization of this potential will require a direct, sustained critique of the individualized consumption practices of late capitalism.

Aglietta's specificity in regard to individualized consumption provides a good starting point for such a critique. He claims that the consumption norm which emerged in twentieth-century American capitalism "is governed by two commodities: the *standardized housing* that is the privileged site of individual consumption; and the *automobile* as the means of transport compatible with the separation of home and workplace."[55] These two commodities are obviously important to the perspective I am developing here, inasmuch as the automobile, alongside its function as a means of transportation between home and work, is also the source of a great variety of consumed convenience; the same can be said of the standardized house, with its array of time and labor-saving devices. I will return to this point of contact between Aglietta's perspective and mine when I focus on transportation technology in a later chapter, but first I must very briefly examine the work of one other contemporary Marxist, Ernest Mandel.

Mandel's *Late Capitalism*, like Aglietta's text, is an elaborate examination of the advanced form of capitalism. And his analysis of modern consumption, like Aglietta's, makes up only a part of the broad scope of his work. So when I focus on Mandel's thoughts on consumption, I must make it clear that I have no pretension of presenting a thorough summation of *Late Capitalism*. The same can be said, of course, for my treatment of Aglietta's text.

On the issue of modern consumption and needs, Mandel can be read as taking a step back from the work of Preteceille, Terrail, and Aglietta. He does not go as far as these other writers in recognizing the influence which the needs of consumers can have on the development of capitalism. Whereas Preteceille, Terrail, and Aglietta ascribe to the needs of workers an important, although non-decisive, role in explaining the proliferation and diversification of the commodities consumed in modernity, Mandel retreats toward the more orthodox Marxist position which underplays the role of such needs.

This is not to say that Mandel is oblivious to the new needs of workers under the semiautomatic, industrial production process. Indeed, he points out that "the substantial increase in the intensity of labour makes a higher level of consumption necessary (among other things, better quality food, greater meat consumption, and so on) if labour power is to be reconstituted at all," and that "the increasing extension of capitalist conurbations lengthens the circulation time between home and work to such an extent that time-saving consumer goods likewise become a condition for the actual reconstitution of this labour power."[56] But these needs of the workers are relatively unimportant for Mandel's explanation of the diversification of consumption in modernity, and he does not develop this issue of needs much beyond the level of the above quotes. He simply notes that such needs are part of an explanation of "the differentiation of the monetarily effective demand of the proletariat in the industrialized countries"[57] and leaves it at that.

As a Marxist, of course, Mandel identifies the capitalist production process as the ultimate source of any developments in needs and consumption. But here too his argument differs from that of the other Marxists I have examined. It is not that the semiautomatic production process, with its new strains and pressures for workers, has brought about needs for new commodities. The development of various time- and labor-saving commodities arises not from the needs of workers (which are ultimately determined by the production process), but rather from the need of capital to find new areas in which to extract profits. From Mandel's perspective, "the basic halmark [sic] of late capitalism" is "the phenomenon of *overcapitalization*, or non-invested surplus capitals."[58] As he explains this phenomenon:

> As long as 'capital' was relatively scarce, it normally concentrated on the direct production of surplus-value in the traditional domains of commodity production. But if capital is gradually accumulated in increasingly abundant quantities, and a substantial part of social capital no longer achieves valorization at all, the new mass of capital will penetrate more and more into areas which are non-productive in the sense that they do not create surplus-value, where it will displace private labor and small enterprise just as inexorably as it did in industrial production 200 or 100 years before.[59]

It was as a result of the pressure of this uninvested surplus capital that various time-consuming household tasks, along with services provided by household laborers, were commodified, if I may use this term. "The housemaid, private cook and private tailor do not produce any surplus-value,"[60] nor does the housewife, who cooks, cleans, and sews for her family.[61] But vacuum cleaners, precooked and preserved foods, sewing machines, and ready-made clothes are all commodities that expand the range of exchange and are produced under the capitalistic wage relation. In other words, such time- and labor-saving commodities are a source of profits which exists beyond "the traditional domains of commodity production."

Mandel's explanation of the expansion and differentiation of consumption, in comparison to those other Marxist accounts examined above, does not appear to be particularly helpful in providing an understanding of the way in which technology has shaped modern needs and consumption. From his perspective, the need for various technological apparatuses in the household is really just a modern form of capital's need for surplus-value; the needs of consumers are of minimal explanatory value. Nonetheless, Mandel's analysis of modern consumption practices does touch upon an issue that is important for understanding the hold which technology has on modernity. Mandel points out that in order for the consumption of the labor force to become diversified, it is necessary that there be a decrease in the portion of laborers' income spent on what he calls the "'pure' means of subsistence." When the "purely physiological" element of workers' consumption decreases in value, then the "historically and socially determined" element can increase.[62] The distinction

Mandel makes here, of course, is of the sort that I am trying to avoid in this text, so to recast his point in the terms I introduced in the previous chapter, in order for workers to overcome the *limits* of the body through the consumption of modern conveniences, they must be able to satisfy the *demands* of the body with only a small part of their wages.

When I introduced this demand/limit distinction, I used the example of ancient Greek agriculture to illustrate the demands of the body, and Mandel is also referring principally to food when he uses the term "pure means of subsistence." In late capitalistic societies, a smaller percentage of income is spent on food than in early (i.e., competitive) capitalistic or precapitalistic societies. This is because "*the age of late capitalism...has been characterized by an even greater increase in labour productivity in agriculture than in industry.*"[63] The "industrialization of agriculture," as Mandel puts it, is another consequence of the "overcapitalization" which characterizes late capitalism. Agriculture is one of those areas into which excess capital flows in its search for profits.

The increased productivity achieved through the mechanization of agriculture has led not only to a decline in the prices of agricultural commodities, but also to a decrease in the number of agricultural workers.[64] So in both these senses (the amount of hourly wages spent for agricultural commodities and the amount of labor hours directly spent in agriculture), Mandel's analysis seems to agree with my earlier claim that modernity is characterized by the relatively small amount of time that it spends satisfying the demands of the body, or at least the body's demand for food (see chapter 2, pp. 35–36).

As I pointed out above, the two consumer items which Aglietta cites as the governing commodities of the modern consumption norm (i.e., standardized housing and the automobile) also fit well with my interpretation of modern consumption practices. By "standardized housing," Aglietta is referring to the prefabricated suburban dwelling, which is built to receive the various appliances which have reduced domestic labor.[65] Electricity is available throughout this house, for lighting as well as for cooking and cleaning appliances. Water for cooking or cleaning is available on demand, as is heat. And sewage disappears in a flush. Such housing, like modern agricultural practices, is geared toward overcoming the limits which the demands of the body place upon the use of time, or the temporal limits of the body.

The automobile, on the other hand, allows people to overcome what I earlier described as the spatial dimension of bodily limits (see chapter 2, pp. 37–38). In the previous chapter, I used the airplane as an example of a technological device that allows people to overcome these spatial limits, largely because air travel clearly portrays the close connection between spatial and temporal limits in the need for ever-faster flights. But Aglietta's example of the automobile has its own particular virtues; the production of automobiles is the model of Fordist production techniques, and the automobile is more clearly a consumer item than is the airplane. And more and more, the automobile is coming to reflect the connection between temporal and spatial limits of the body. I have in mind here the drive-through windows of fast-food restaurants and the tendency to extend such facilities to banks, liquor stores, grocery stores,and so on. In these situations, the automobile allows people to overcome the limit of distance as well as save time in various daily routines.

But if these Marxist accounts of modern consumption end up so close to the perspective I am trying to establish, why have I set them up as a hindrance to an understanding of modern technology? Why spend the first part of this chapter criticizing the Marxist emphasis on production when writers like Aglietta and Mandel ultimately identify certain features of modern consumption which are important to my perspective? The reason I have been critical of these Marxists is that their preoccupation with the capitalist production process results in a certain narrowness in the historical and material dimensions of their perspective. Early in this chapter I distinguished these two elements of historical materialism and indicated that Preteceille and Terrail employ them as safeguards to keep their analysis of modern consumption from falling back into the idealism of liberal economic thought. I made this distinction in order to question the Marxist doctrine that the ultimate historical determinant is the production process (the materialist dimension) and to suggest that this doctrine may have become a hindrance to understanding modern technological consumption, if not late capitalism itself. At this point, I can further specify that early criticism.

When Aglietta and Mandel identify modern consumption practices, which revolve around inexpensive and readily available agricultural commodities, automobiles, and technologically organized housing, they have in mind changes which occurred primarily after the Second World War. According to Mandel, the great advances in agricultural productivity which caused food prices to decline, thereby freeing up the income of workers for more diversified consumption, occurred in this period.[66] And the social norm of consumption which is central to Aglietta's analysis of Fordism emerged in the 1920s but flourished after the war.[67]

From my perspective, however, the important shifts in consumption practices occurred much earlier, as early as the eighteenth century, and are not simply the effects of developments in production techniques. The intensification of consumption which characterized the 1950s can be explained, in part, by twentieth-century developments of the production process, and Aglietta's and Mandel's analyses are helpful in that regard. But when it comes to the features of the actual commodities consumed, one's analysis has to expand beyond the twentieth century. Conveniences in agriculture, transportation, and other household activities were developed, produced, and consumed in the nineteenth century, as will be seen shortly. But it is not only the historical dimension of the Marxist interpretation of modern consumption which must be expanded; the materialist dimension must be expanded as well.

The materialism of Marxist analysis is limited to the capitalist production process. Most Marxists are unaware of any other material considerations which may be helpful in understanding certain features of modernity. What I have in mind at this point as such an other material consideration is the vast amount of unsettled land that was available in America throughout the eighteenth and nineteenth centuries. (Another material consideration, also overlooked by Marxists, will be examined in a later chapter.) This spatial condition, I will argue in the next chapter, played an important role in the development of the consumption pattern or norm which has been identified with the United States and has spread throughout much of the world. So although I do not ascribe ultimate determination of consumptive activity to the production process, my approach is nonetheless materialistic, inasmuch as land and space are material considerations.

CHAPTER 4

Settling American Space

MARX'S INSIGHT

In fairness to Aglietta, I must mention at the outset that he does discuss the American frontier experience in *A Theory of Capitalist Regulation*, but for him the ideological value of the frontier is predominant. To quote Aglietta:

> The frontier principle was more than is implied simply by its literal content, in other words the mere domestication of a geographical space. It was rather an ideological principle expressing the ability of the American nation to polarize individual activity in a direction of progress. Indeed the industrial bourgeoisie was later able to get the whole of the nation to accept the technological trans-formations induced by relative surplus-value by presenting these as the building of a 'new frontier'.[1]

There is little doubt that the "frontier principle" has served this ideological purpose, and continues to serve it as the United States races, and may perhaps cooperate, with other nations to colonize outer space. But there is a material dimension of the American frontier worthy of attention. Aglietta does not totally ignore this dimension, but for him the abundance of land in the United States "enormously favored" the creation of a surplus of agricultural commodities, one of the preconditions for capitalist industrial production.[2] So from Aglietta's perspective, the frontier, in both its ideological and material dimensions, served the development of capitalism in America.

My interpretation of the American frontier is quite different from Aglietta's. To begin with, I emphasize the material dimension of the frontier, the "mere domestication of a geographical space." Furthermore, I view the frontier not as a boon, but a threat to capitalism, at least initially. Ironically, this non-Marxist perspective is supported by certain insights of Marx himself. The eighth and final part of the first volume of *Capital*, concerning "The So-Called Primitive Accumulation," is described by Aglietta as "decisive,"[3] and I would agree, although apparently for different reasons. (Aglietta does not expand on this point.) In this section of *Capital*, Marx examines the manner in which feudal society was transformed by and for the forces of capital. His concern lies less with the eclipse of restrictive feudal relationships (which bourgeois, or liberal, theorists emphasize) than with changes in material conditions. In order for capitalism to flourish, a supply of laborers was needed which was free not only from the obligations of serfdom, but also, and more importantly, free from any attachment to the land. Marx stresses this point:

> In the history of primitive accumulation, all revolutions are epoch-making that act as levers for the capitalist class in course of formation; but, above all, those moments when great masses of men are suddenly and forcibly torn from their means of subsistence, and hurled as free and 'unattached' proletarians on the labour market. The expropriation of the agricultural producer, the peasant, from the soil, is the basis of the whole process.[4]

This is the point I was referring to in the beginning of the previous chapter when I claimed that there exists in Marx's thought an often overlooked appreciation for the transformation in consumption practices which accompanies capitalism. The expropriation of the peasant from the land destroys the peasant's ability to produce for household consumption; the peasant is torn from the means of subsistence, not just the means of production. Marx recognizes that this process of expropriation took different forms in different countries, but he identifies the enclosure of the commons in England as the classic form.[5]

The enclosure of the English commons began in the fifteenth century but became widespread during the eighteenth.[6] Prior to enclosure, the peasant who owned or rented a cottage on a manor enjoyed several rights of common which included the right to grow strips or rows of crops in the arable fields of the manor. On the remaining land of the manor, called the waste, commoners had the right to pasture as much livestock as was required to cultivate their strips, cut hay for winter feeding of the livestock, collect timber for building and repairing agricultural implements, cut peat or turf for fuel, and if there were streams and ponds on the waste, to catch fish. These and other rights of common were defined by, and limited to, the needs of the household. These rights were also frequently exercised by squatters who lived on or near the manor but had no legitimate claim for such exercise.[7]

With the enclosure of the common arable and waste lands came the extinction of the various rights of common, and the extinction of the small farmers who were able to satisfy the needs of their households through their own productive activity. As compensation for the loss of these rights, commoners were given a small parcel of land as an element of the act of enclosure.[8] But these small plots were insufficient to provide for the needs of the household, and often the costs of ditching and fencing the allotment (which were also required by the acts of enclosure) were so high that the owner had no choice but to sell the land to a larger landowner, and join the ranks of agricultural or manufacturing laborers.[9] Enclosure also had the effect of rendering the laboring class completely dependent on the wage relationship with the employer, thereby making laborers more docile and regular. As one proponent of enclosure put it at the end of the eighteenth century: "'The use of common land by labourers operates upon the mind as a sort of independence'. When the commons are enclosed 'the labourers will work every day in the year, their children will be put out to labour early', and 'that subordination of the lower ranks of society which in the present times is so much wanted, would be thereby considerably secured.'"[10]

In contrast to this situation in England, which was crucial to Marx's understanding of capitalism's development, the United States had vast amounts of unsettled land throughout the nineteenth century. As English peasants were forced off the land, and lost the ability to provide for their needs on their own, Americans were moving West and bringing virgin forest, and eventually the prairie, under cultivation. This difference, and the important consequences of it, were not totally lost on Marx. In the last chapter of the first volume of *Capital*, "The Modern Theory of Colonization," he discusses this difference. In 1866 Marx claimed that, "speaking economically," the United States belongs to the category of "real Colonies, virgin soils, colonized by free immigrants."[11]

What Marx finds interesting about such colonies is that they give the lie to arguments of liberal political economists.

Such economists, claims Marx, confuse "on principle two very different kinds of private property, of which one rests on the producers' own labour, the other on the employment of the labour of others."[12] In interpreting the capitalist economies of Western Europe, the political economist "applies the notions of law and of property inherited from a pre-capitalistic world"—that is, one where producers own the means and the products of their labor—and forget that capital, which is based on the labor of others, "not only is the direct antithesis of the former, but absolutely grows on its tomb only."[13] This confusion is cleared up for the political economists once they confront the situation in the colonies.

In the colonies, the fundamental antagonism between precapitalistic property and capital becomes obvious as the ideological foundations of capitalism come to be experienced as reality. The availability of unsettled land provided laborers with an alternative to the wage relation, thereby making the relation between employers and laborers more truly a contract since workers were free to refrain from entering into it. And the liberal conception that property rights are grounded in the natural right individuals have in their bodies and the labor of their bodies[14] becomes a reality as laborers abandon the wage relation and cultivate their own property for themselves. The effects of this unfortunate concordance between ideology and reality, at least as viewed from the perspective of capital, are described by Marx as follows:

> This constant transformation of the wage-labourers into independent producers, who work for themselves instead of the capitalistic gentry, reacts in its turn very perversely on the conditions of the labour market. Not only does the degree of exploitation of the wage-labourer remain indecently low. The wage-labourer loses into the bargain, along with the relation of dependence, also the sentiment of dependence on the abstemious capitalist.[15]

In this chapter on modern colonization, Marx cites E. G. Wakefield's *England and America*, which bemoans the fact that in colonies with an abundance of land, "'the supply of labour is always, not only small, but uncertain.'"[16]

Marx, of course, did not overestimate the threat which open space posed to capitalism, and he undoubtedly understood that capital was up to this challenge. He noted Wakefield's plan for "systematic colonization," whereby land prices would be raised to a point at which wage-laborers would have to save for years to be able to afford a sufficient piece of property, and the surplus funds from these land sales would then be used by the government of the colony to import replacement laborers from Europe.[17] But in the case of the United States, Marx recognized that such a scheme was being rendered unnecessary by several developments. The successive waves of immigrants provided eastern industry with a reserve of dependent laborers; the government's debt from the Civil War would call for heavy taxation; and capital was being increasingly consolidated in mining and railroad companies. All of this indicated to Marx that "the great republic" had "ceased to be the promised land for emigrant labourers."[18]

SETTING AMERICA'S SPACE IN ORDER

Despite his insights into the problems which unlimited space posed for capitalism, Marx did not note the irony, in the case of the American colony, of plans such as Wakefield's. In 1862, five years before the publication of the first volume of *Capital*, the government of the United States began to give away 160-acre homesteads to settlers who lived on and improved the land for five years. Obviously, the passage of the Homestead Act did not mark the victory of a peasant economy over capitalism, so how can this free land policy be explained? It may be tempting to say of this what Marx said of the British attempts to legally restrict the enclosure of the commons and retain small agricultural plots for cottagers (i.e., laws of tillage): those who supported such laws had not yet come to realize that the wealth of the nation and the poverty of the people were complementary.[19] Perhaps the generous American land policy of the late nineteenth century was similarly anomalous; Congress had not yet caught on to this principle of modern statecraft and so was not using its vast holdings of land to increase the national wealth (i.e., capital).

There are problems with such an explanation, however. First, given this explanation, one would expect the federal government to have learned its lesson quickly in the face of its enormous Civil War debt and begun to sell, not give away, the public domain. But this was not the case, and the Homestead Act remained in effect into the twentieth century. Second, this explanation does not account for the fact that the government had already learned this lesson about the national wealth in the eighteenth century and had initiated a program for selling public land in order to pay the Revolutionary War debt.[20] So the Homestead Act cannot be dismissed simply as a mistake made by an inexperienced government. On the contrary, the Homestead Act marks the culmination of a long trend in the land policy of the United States away from its conservative beginnings as a means for raising revenue toward its more 'democratic' role as promoter and protector of family farms. So the question remains: how could a nation emerging as the leader in industrial capitalism, afford to give away land to small homesteaders? How could a country with an expanding capitalist economy undertake a policy of promoting that form of property which was the antithesis of capital and, according to Marx, upon whose tomb only capital grows?

The answer I would like to offer to these questions is that the threat which was posed to capitalism in America by the vast amount of unsettled land (unsettled, at least, by white men) had been largely eliminated by the 1860s. The manner in which the threat of American space was neutralized provides insights into the later development of capitalism in America, especially in regard to the American standard of consumption. My claim is that the American fetish for technological conveniences can be understood, in part, as an outcome of the subjugation of American space. To support this claim, it is necessary to discuss not only American developments in transportation and farm machinery, but American land policy as well. But to begin with, I must be a little more specific about the nature of the threat.

The problem posed by America's vast amount of unsettled land was felt even in the prerevolutionary period, although not so much as a threat to the labor supply as a threat to the lives of frontier settlers. White settlers continuously encroached upon territory that the colonial assemblies and the British Crown recognized as belonging to the Indians. In an attempt to quell this practice, a practice which engendered much hostility among the Indian population, George III's Proclamation of 1763 forbade his subjects from settling west of the Appalachian mountains.[21] But this measure and various other attempts by colonial governments to restrict the settlement of unoccupied land were of no avail. In 1768, for example, the assembly of Pennsylvania passed an act requiring settlers on the forbidden land to remove upon penalty of death; but by 1775, there were between 25,000 and 30,000 settlers beyond the Appalachians.[22]

Even within the colonies themselves, settlers were quick to establish homesteads on unoccupied land that had been granted to certain proprietors by the Crown.[23] The proprietors of Pennsylvania, Maryland, New Jersey, and the Carolinas sought to extract from these settlers a form of feudal obligation known as quitrent, which was a single payment made in lieu of all other feudal obligations. However, these attempts to bridle the clearers and cultivators of the virgin forests along the Atlantic seaboard were unsuccessful and often met with open revolt.[24]

This practice of settling on unoccupied land, which eventually came to be known as squatting, continued after the American Revolution, and squatters had as little respect for the authority of the government of the United States as they had for the British Crown, at least when it came to attempts to restrict their settlement on unoccupied land. After the Revolutionary War, several states eventually ceded to the federal government the large tracts of land that had been granted to them by the British Crown.[25] As I have mentioned, the Continental Congress sought to relieve the federal government's war debt through the sale of these lands. In one of the first and most important acts of the Continental Congress, an ordinance was passed in 1785 which established a program for selling the land northwest of the Ohio River (the Northwest Territory), which had been ceded to the federal government by Virginia in 1784. To deal with the problem of squatters who had settled on the most promising land in this territory,[26] the ordinance granted the Secretary of War the authority to use federal troops to remove these trespassers.[27]

This policy of removing squatters by force was largely ineffectual, due both to the overwhelming number of squatters who settled illegally on government and Indian land and to the tenacity of the squatters, who often returned to rebuild their destroyed homesteads after the troops had left.[28] Nevertheless, the policy of removing squatters from government land remained in effect until 1841 (with some important exceptions which will be mentioned shortly). And federal troops were obligated to remove American trespassers from Indian land according to the terms of the various nineteenth-century treaties between Indian tribes and the government of the United States.[29]

Although the military response to the problem of squatters was a failure, other elements of the Ordinance of 1785 were ultimately more successful in undermining the threat posed by open space to the formation of capital. It must be emphasized here that the problem or threat of unsettled land was not simply that squatters might establish claims without paying for them; even if the government was able to exact payment from every family that settled on the public domain, there was still the problem of independent homesteaders who would be beyond the control of capital. Prohibiting 'free' settlement on the public domain, in the sense of nonpayment, was only part of the problem; the freedom that could be attained through the lawful purchase of land was the more far-reaching threat. The response to this facet of the threat of space is more varied and subtle than the violent treatment of squatters, and it is also more closely related to the theme of this text.

To begin with, the Continental Congress thought it necessary to regulate the pattern of settlement as it moved westward, and toward this end, the Ordinance of 1785 required that the land in a given area be surveyed before being offered for sale.[30] And the particular surveying system which was adopted by the Continental Congress further indicates its serious concern for controlling the westward movement of settlers. According to this surveying system, the vastness of American space was to be dissected into equally sized squares in parallel rows. Starting from the point where the Ohio River crossed the Pennsylvania border, a line was to be surveyed due west. Similar base lines were to be established every six miles south of this line, up to the border of the Ohio. These parallel base lines were in turn crossed by north-south meridians, which were also drawn at six-mile intervals. Each six-mile by six-mile square formed by these intersecting lines was a 'township,' and these were further subdivided into 36 one-square mile 'sections,' of 640 acres each.[31]

The first survey accomplished under this system was to consist of seven north-south rows of townships, called 'ranges,' running south from the original east-west base line, down to the Ohio River. Once the survey was complete, the land was to be sold either by township or section, that is, either in 36-square-mile or one-square-mile lots.[32] By limiting sales to surveyed land only, and by requiring that entire townships or sections be bought, the government sought to reap the full value of its land holdings. Once a surveyed area had largely been sold, the surveying of a new region would proceed. This system allowed the government to ensure that all land was developed, not just the most desirable land.[33] The constitutional government created in 1789 retained this system of land survey and sale and used it to regulate the settlement of the various lands purchased by or ceded to that government.

The rectilinear system of land surveying stands in sharp contrast to the less 'rational' survey system which prevailed in southern states such as Kentucky and Virginia, where feudal impulses were strongest.[34] Under that premodern system, settlers were free to chose the most desirable unclaimed land, and to survey it in a haphazard manner which relied on natural, and therefore changeable, landmarks, such as streams and trees. Aside from the boundary quarrels that such a loose arrangement generated among settlers, it also had the disadvantage, from a modern perspective, of allowing certain areas to remain 'unimproved,' or undeveloped. The presence of such 'waste' was one of the obstacles England faced in its modernization, and the enclosure movement was in part a response to this problem. Swamps, marshes, and other 'useless' lands were reclaimed through the enclosure process in the sixteenth and seventeenth centuries and were then used for either pasture or crop cultivation.[35] The survey system employed by the United States prevented the accumulation of such waste land in the first place, by setting everything in order before legal settlement was permitted.[36] This American attempt to set its space in order, to establish a grid of borders, displays an ordering tendency that has been cited as a birthmark of modernity.[37]

But to remain focused on a more narrow examination of American space, the survey and sale system adopted in the Ordinance of 1785 helped not only to ensure that all land was developed, but in conjunction with a restrictive pricing mechanism, it also helped to undermine the threat of independent homesteaders. As previously mentioned, the surveyed land was sold either in townships to speculators or in sections to homesteaders and speculators; nothing smaller than a 640-acre section could be purchased from the government. The price per acre for these surveyed parcels was to be established by competitive bidding at auction, but the minimum price per acre was set at one dollar. So even for the least desirable section in a surveyed area, the homesteader would have to pay a minimum of 640 dollars, plus a surveying fee of one dollar per section.[38]

The price of one dollar per acre may today appear to be a very generous gesture on behalf of the government toward settlers, but state-owned lands were being sold at that time for significantly lower prices.[39] So at the outset, the government of the American 'colony' had developed a land regulation system which, at least in its pricing mechanism, closely resembled the price-fixing plan suggested by Wakefield. Prices were raised to a level which prevented the widespread settlement of newly surveyed land by individual homesteading families. The $641 minimum price placed homesteading beyond the means of many,[40] and speculators could easily raise the price level beyond the means of other would-be settlers during the auction.[41]

It was not only the price that prevented many from acquiring homesteads; the minimum purchase size was also a deterrent. Those who had earlier settled the Appalachian Plateau, just west of the Carolinas and Virginia, usually established very small clearings of four to five acres, on which they planted the food that would feed their families. The size of these farms increased gradually over the years, as the families did, and might occasionally reach a size of four hundred acres.[42] But after 1785, any settler who legally purchased a homestead from the government had to begin with 640 acres. This is not to say that all of the purchase had to be cleared and planted immediately, but because the settler also had to pay at least 641 dollars for the land, which was quite likely borrowed at interest, there was some pressure to clear the land quickly and produce surpluses in order to repay the investment or loan which had been undertaken. Of course, smaller homesteads could be purchased from speculators, but at a significantly higher price than the government minimum.

Not surprisingly, therefore, the early pace of land sales in the Northwest Territory was not very brisk.[43] This was due in part, no doubt, to the resistance the Indians offered to white settlement, but the land policy established by the Continental Congress in 1785 was also a hindrance, at least to the small homesteaders. In 1796 the policy was revised so that it would be easier to acquire a homestead, but this was not accomplished by lowering the minimum sale size or the minimum price per acre; on the contrary, the minimum size was maintained at a full section, and the minimum price was doubled to two dollars per acre. What made it easier for the individual homesteader to buy land was the introduction of a credit system which allowed the settler to pay for the land over a four-year period.[44] But with the full-section minimum, this policy still favored the speculator over the homesteader. It was not until 1800, when the minimum size was reduced to a half-section (320 acres) and the credit system was refined that land sales to homesteaders flourished.[45] The relatively easy terms of credit required a 5 percent initial payment (thirty-two dollars minimum), 25 percent payment after forty days, and another quarter-payment at the end of the second, third, and fourth years. The rate of interest on the balance was 6 percent.[46] In 1804, the government reduced the minimum size to a quarter-section, making it even easier for settlers to acquire their homesteads on credit.[47]

While the availability of credit for land sales is often interpreted as a government concession to the small homesteaders, who otherwise would have been unable to buy land, this policy can also be read as a subtle way of undermining the threat independent farmers posed to the American economy. Under the terms of the Land Act of 1800, small homesteads did multiply, but the owners of these lands were debtors to the federal government. These farmers had little choice but to produce cash crops which could feed the urban population, so that they would be able to make the payments on their land. Self-sufficient farming was not a possibility for these debtor farmers; they had to become engaged in the capitalist exchange economy, both as suppliers of foodstuffs and as consumers of industrially manufactured items which they no longer had time to produce for themselves. (Of course, the availability of such manufactured items and the ability to sell crops for cash depended on crucial developments in transportation. My focus here, however, is on the way space was brought to order, or subdued, by American land policy. I will get to the way people and things were set in motion, or how the limit of distance was overcome, in the following chapter.)

The emergence of government-sponsored credit at the end of the eighteenth century is a significant development, since credit has become one of the primary features of contemporary consumption. Both Aglietta and Mandel indicate the important role credit plays in fostering the current mode of consumption. But both of these theorists treat widespread consumption on credit as a twentieth-century phenomenon. From their perspectives, consumer credit became an important feature of capitalism only when the intensification of the production process in the 1920s demanded the close integration of production and consumption.[48] The point I want to stress is that buying on credit was an important phenomenon at the beginning of the nineteenth century in regard to land purchases. As one land historian put it, credit was "the very life blood of the West."[49] From my perspective, the government's extension of credit to settlers was a way of undermining the threat that the vast amount of unsettled land posed to capitalism in the United States. Credit emerged not as a development of late capitalism, but as capital's response to a material condition other than the capitalist production process.

The extension of credit to homesteaders, therefore, appears to have been part of a shift in the land policy of the United States between the Ordinance of 1785 and the Land Acts of 1800 and 1804. The initial policy tried both to restrict the settlement of the public domain (by prohibiting squatting and evicting squatters, by requiring that surveys of entire regions be complete before sales could begin, and by establishing a prohibitively large minimum lot size), and to maximize the revenue generated by land sales (by selling the public domain at auction, and by establishing a minimum price of one dollar per acre, and then doubling it). In 1800, however, the restrictive character of the land policy seemed to give way to government efforts to promote settlement. The reduction in minimum lot size from 640 to 320 to 160 acres, plus the provision of credit, sparked a surge in land sales.[50]

Not only did the land policies of the early nineteenth century foster, rather than restrict, settlement, but they did so at the expense of the other purpose of the original policies—the raising of revenue for the federal government. In fact, the liberal credit terms of the Land Act of 1800, which remained in effect until 1820, caused an explosion in the debt owed to the federal government, very little of which was actually repaid. By 1819, the government had sold land worth 44 million dollars, but had received payment for only about half, and at that point the situation seemed only to be getting worse. Between 1815 and 1819, the amount owed to the government ballooned from three million to seventeen million dollars, and the government had to repeatedly pass relief legislation for its many debtors and extend the terms of their payments.[51]

The combination of easy credit and widespread speculation in land led to a continuing inflation of land prices. In 1819, this bubble burst and a panic ensued. As a consequence, the government abolished its credit system in 1820 and required that all future land sales were to be paid for in cash at the time of purchase.[52] Although the abolition of credit did make it harder for the settler to acquire land, the minimum price and size requirements were both reduced, from $2 to $1.25 per acre, and from 160 to 80 acres.[53] So, despite the tremendous debt owed to the federal government, land policies continued to move away from their original, conservative form toward what appears to be a more populist model.

It was not simply the credit system of 1800 which undermined the profitability of government land sales, however. From the outset, the practice of selling land by public auction had been manipulated by speculators to their own advantage. Speculators often made arrangements before auctions whereby they would determine who would buy what piece of property and then agree not to bid against one another. The result of this collusion, of course, was that—prior to the introduction of credit, which allowed homesteaders to join in the bidding—land prices were not raised much above the minimum at auction.[54] So even before the credit explosion of the first two decades of the nineteenth century, the federal government was not receiving the revenue expected from its land sales.

With the abolition of the credit system, homesteaders lost what little advantage they had gained against the speculator, and land sales to homesteaders dwindled in the 1820s. This is not to say that settlement dwindled, however. Squatters settled on the public domain in the hope that they would be able to afford their eighty-acre claim when the surveys were completed and the auctions held.[55] These squatters, of course, were easily victimized by speculators, who would often agree, for a fee, to refrain from bidding against the settlers. The squatter might get the land at the minimum price, but only after paying off the threatening speculators.[56] And if the speculator wanted the land, not just a payoff, he could easily outbid the settler and take the homestead.

Another method commonly used by speculators to undermine the independence of squatters, was to lend them money for the purchase of their claims—at exorbitant interest rates, of course. Such lending, claims one historian, "proved to be one of the most lucrative business opportunities on the frontier."[57] Here too, the speculator skirted the law, but in this case he was avoiding state and territorial laws against usury, not federal land laws. While the usury laws limited interest to 10 or 12 percent, speculators entered the squatter's claim in the name of the speculator, and bonded the squatter to pay a specified amount by a specified date, in order to receive the title. Through this technique, the speculators were able to collect what was, in effect, interest at rates as high as 100 percent.[58]

So although the U.S. land policies in the early nineteenth century appear to have been moving in the opposite direction from Wakefield's plan for "systematic colonization," the effect of these policies was the same. Land prices, directly or indirectly, were raised to a level which made it difficult for independent homesteaders to acquire property. The profit from this policy, however, went not to the government, as Wakefield had suggested, but to private speculators. In either case, the threat of an independent agricultural population was undermined.

In order to avoid painting too pathetic a picture of the vulnerable, hapless squatter, I must mention that as squatting increased in the 1820s and thereafter, squatters learned the lessons presented by the practices of the hated speculators. They adopted the speculators' technique of uniting in an organization which would allow them to prevent land prices from rising at auction. These "claim clubs" or "claim associations," as they have come to be known, sprang up throughout the frontier.[59] Where claim clubs were established, therefore, auctions were a sham, just as they were where speculators organized themselves, and land prices were held down to the established minimum.[60]

And to avoid painting too noble a picture of these organized squatters, I must also point out that many members of the claim clubs were not merely protecting their right to that with which they had mixed their labor, to paraphrase Locke. Often squatters were speculators themselves. By acquiring de facto title to their claim through the claim club, those squatters who were unable or unwilling to pay for their land, even at the minimum price, could sell their title to another before the auction. The purchaser of such a title, of course, also had to buy the land again at the auction.[61] And these speculating squatters sometimes claimed tracts of land which were larger than the amount they could cultivate, and used the force of the claim club to keep other squatters off their extensive holdings.[62] So much for the noble squatter.

In any case, the primary target of the claim clubs was not other settlers, but wealthy speculators, whom they would prevent from bidding through the threat (or use) of violence.[63] But after 1830, the squatters were no longer on their own in their struggle against speculators. Throughout the decade of the 1830s, Congress repeatedly passed Pre-emption Acts, which allowed those who had settled on the public domain and had made specified improvements on the land to buy up to 160 acres of such land at the minimum price of $1.25 per acre.[64] As a result of these acts, settlers no longer had to vie with the speculators at auction and were no longer susceptible to their extorsive tactics.[65]

The Pre-emption Acts of the 1830s applied only to land that had already been improved prior to the act. In other words, these acts were retrospective and, in a sense, they forgave past transgressions by those who illegally settled on public land. In 1841, however, Congress passed a Pre-emption Act which was prospective. This act extended the terms of the earlier acts to all those who would settle and improve the public domain.[66] Squatting was no longer illegal.[67] The government finally seemed to recognize that the squatters had some right to the land they had improved.

This right, of course, was not the precapitalist property right espoused by political economists like Locke (see Chapter 4, note 12). By the pre-emption process, squatters only gained the right to *buy* the land they improved through their labor. But in 1862, with the passage of the Homestead Act, something very close to that precapitalist property right was established. Under the terms of this act, any man at least twenty-one years of age, the head of a household of any age, or a widow, could gain title to any surveyed public land up to 160 acres, which he or she lived upon and improved for a five-year period.[68] Through their labor on the land (or the labor of someone in their hire, as Locke intended), settlers gained a property right in that land. As I noted at the outset of this discussion of American space, there is an element of irony in the passage of the Homestead Act at the very moment when industrial capitalism was expanding in the United States. There is such irony, at least, if one takes Marx's thought into account.[69] But at this point in my discussion, the irony should have begun to dispel.

To summarize, in the development of American land policy from 1785 to 1862, the shift from its conservative origin to its populist conclusion was accompanied at every turn by widespread speculation. Every policy revision which favored the small homesteader was turned by speculators to their own advantage. The reductions in price and size limits and the provision of credit all helped the speculator as well as the settler. And the Pre-emption Acts, which forgave and ultimately legalized squatting, were fraudulently abused by speculators. Even the Homestead Act was a boon to speculators. According to the terms of the Act, a settler could commute the five-year residence requirement after six months upon payment of 1.25 dollars per acre. After 1880, speculators began to fraudulently take advantage of this provision to increase their land holdings.[70] Since speculators had been able all along to acquire vast holdings of desirable land or, alternately, to extort money from settlers, it is not very surprising that the United States could afford in the middle decades of the nineteenth century to legalize squatting and then to give land away. The best farm land was largely owned or occupied by 1862, and most of the remainder of the public domain was unfit for agriculture.[71] If a settler wanted good farm land, he or she still had to pay the speculator's price.

AGRICULTURE AS A LIMIT OF THE BODY

But aside from the land policies of the U.S. government and the speculators who flourished under those policies, there were other forces operating in the first two-thirds of the nineteenth century which helped to undermine the threat of American space. I have already touched on one of these forces in my discussion of Mandel's *Late Capitalism* when I mentioned the "industrialization of agriculture," which Mandel treats as a feature of the advanced capitalism of the twentieth century (see Chapter 3, p. 61). To reiterate Mandel's point, the industrialization of agriculture was a result of overcapitalization; agriculture was one of the areas into which surplus profits flowed after the "traditional domains of commodity production" were thoroughly industrialized.

The point I want to make is that the industrialization of agriculture—or perhaps 'mechanization' is a better term—occurred in the nineteenth century and was actually well-established by the time the Homestead Act was passed. Furthermore, one can interpret this process not simply as the result of advanced developments in capitalism, as Mandel does, but as an attempt to undermine the threat of unsettled space. In fact, the mechanization of agriculture is closely bound up with the restrictions which were imposed by U.S. land policy.

As previously mentioned, the minimum lot sizes established by the various land acts, along with the minimum prices, had an effect on the way in which the land was settled. Those who sought to legally settle on the public domain in the nineteenth century could not afford to clear and cultivate the land at the pace common in the eighteenth century. A small clearing which provided food for the family could no longer be gradually expanded so that after several years it would produce commercially.[72] If a settler purchased land from the government, he or she had to buy at least the minimum acreage at the minimum price. A smaller homestead could be bought from a speculator, but at a much higher price. In either case, the settler most likely bought on credit, either from the government or the speculator, and had to begin making payment soon thereafter. It was imperative that the settler clear and plant as much land as possible as quickly as possible. The objective of such settlers was not so much to produce food to satisfy the demands of the family's bodies, as it was to produce surpluses which could be used to pay off land debts. Farming had become less the satisfaction of a demand of the body than the practice of overcoming limits which hinder the production of food as a commodity.

The first limit to be overcome was the clearing of the land, and in almost every area of the country, the axe was essential to this task.[73] If it was not needed to clear the enormous trees which faced the earliest settlers who headed west into the virgin forests of the Appalachian Mountains, it was needed to produce the implements necessary to farm the plains and to construct fences and buildings. Consequently, the axe underwent a complete transformation in America. The broadaxe, which had remained unchanged for centuries in Europe, was made broader and straighter in the blade, heavier in the head, and lighter and more flexible in the handle. Felling trees was accomplished much more quickly and with less effort by the American, or Kentucky, axe than by its European predecessor.[74]

Various other types of axes were also developed, each capable of accomplishing its particular task (for example, bark-stripping or rail-splitting) in the quickest and easiest manner. These various tools had reached standardized form by the mid-nineteenth century and were being mass-produced.[75] Although the technical development of the axe appears to be quite modest, it does reflect the early form of the American tendency to produce the most convenient tool for the task at hand. And these axes were convenient not just in the premodern sense of being appropriate to the task; they were also designed to promote the ease and comfort of the user and were therefore convenient in the modern sense of this word as well. (See Chapter 2, pp. 39–40 for the discussion of these two senses of 'convenience'.)

The other tool which was essential to clearing the land was the plow. Well into the nineteenth century, this tool was produced by the farmer and was made of wood.[76] The problem with wooden plows, of course, was that they often needed to be repaired, especially in areas with heavy soil and in the prairies, where the wooden share, or cutting prong, was no match for the extremely thick sod. With the changes in land settlement that occurred in the nineteenth century, the settler could ill afford to spend time refashioning and replacing the broken or worn-out parts of the plow. Around the turn of the century, attempts were made at constructing plows out of cast iron, and by 1819 plows with durable, replaceable iron parts were being produced commercially.[77] By 1850, tens of thousands of plows of over one hundred different varieties were being produced each year in cities such as Worcester, Massachusetts and Pittsburgh.[78] And in the 1850s John Deere was successfully producing plows made of cast steel, which were strong enough to handle the prairie sod without needing frequent repair and smooth enough so that the soil no longer had to be periodically scraped from the plow, as was the case with iron plows.[79]

As a result of these technical developments in the plow, settlers were able to clear much larger areas much more quickly than with the wooden model. Siegfried Giedion, in noting the contrast between the settlement of the plains of Europe and Asia and those of North America, raises the point which I would like to stress:

> Other great plains had been brought under the plow. But the opening of the Russian plains and of the vast tracts of China extended over centuries. Compared to these the development of the Middle West took place within a few decades, almost *by elimination of the time factor.*[80] (Emphasis added.)

This concern for eliminating the time factor, or overcoming the limits on time imposed by the task of clearing the land, arose in part from the way in which the space of America was dissected and sold to settlers. And, of course, it was not just the clearing of land that had to speeded up. The planting of seeds and the harvesting of crops had to be quickened as well if the settler was going to be able to take full advantage of the holding and pay off his or her land debt. By 1860, many devices designed to quicken the various facets of crop production were in widespread use.

Most of these developments were centered on wheat, the primary commercial crop of the Midwest.[81] The other important crop was corn, the principal grain consumed by the farm household, including the livestock.[82]

Seed drills, which uniformly plant and cover seeds, were first developed in England in the eighteenth century,[83] but they attained a workable, efficient form in the United States between 1840 and 1860.[84] The use of seed drills, which were drawn along by horses, not only accelerated the seeding process, but made it more uniform. Hand-cranked broadcast seeders had already been developed, and although these were faster than sowing by hand, they still scattered the seed in every direction. Such broadcasting may have been acceptable for wheat seed, but it was not for corn, which had to be weeded, or cultivated, periodically. Drills planted the seeds in straight rows with uniform spacing, which allowed the farmer to cultivate the soil between rows of corn with a horse-drawn hoe, which again was much faster than hoeing by hand. In any case, by 1860 most wheat and much corn was planted by the seed drill,[85] which cost around $125.[86]

Another very important development in farm machinery also occurred about the same time that the seed drill was being perfected. The traditional method of harvesting wheat was to cut it by hand using a sickle or a cradle, the latter being larger and more cumbersome, but also faster, than the sickle. With a sickle, a farmer could harvest one-half to three-quarters of an acre in a day, and with a cradle, two to three acres could be cut.[87] Unlike crops such as corn, wheat has to be harvested within a narrow space of time. As long as farmers had to rely on hand tools for harvesting, their income from wheat was severely limited, and any extension depended on an uncertain and expensive supply of migratory labor to help at harvest.[88]

In 1834, however, Cyrus McCormick patented a mechanical reaper which could harvest up to twelve acres in a day with the assistance of two workers—a twofold improvement over the cradle.[89] Other reapers had been patented before McCormick's, both in the United States and England, but his was the first to gain widespread use. After years of contracting other firms to build the reapers, in 1848 McCormick opened a main plant in Chicago.[90] He began with thirty-three employees and by 1849 had one hundred twenty. In light of Aglietta's and Mandel's emphasis on the production process of the twentieth century, it is interesting to note that McCormick's factory had features that are usually considered twentieth-century innovations. The Chicago factory had a central source of power—a steam engine—that drove fourteen or fifteen machines, such as metal lathes, and the material was moved on a rail system. After this factory burned in 1851, the new factory that was built had a conveyor system and used automatic-feed machinery.[91]

In regard to farm practices, McCormick's reaper and the thirty or so competitors that emerged in the mid-nineteenth century, revolutionized the production of grain. A mechanized rake was added to perform the task of raking the cut wheat off of the reaper and laying it in a row on the ground.[92] This addition doubled the efficiency of the reaper, since one of the two laborers had previously performed that task. By 1869, approximately eighty thousand such reapers had been sold, and the average price was $125.[93] The widespread use of the mechanical reaper was a major reason, along with increased land sales under the Pre-emption Act, for the 72 percent increase in wheat production that occurred in the United States in the 1850s.[94]

By 1880, the harvesting of wheat had been completely mechanized as the last time-consuming task involved in wheat production was brought up to the pace required by nineteenth century American agriculture. The binding of the grain into bundles or sheaves in which it would dry had been traditionally performed by hand as the grain lay on the ground. Around 1870, a conveyor was added to the reaper which carried the cut grain up to a table, where the wheat could be bound, although still by hand.[95] But the binder no longer had to walk through the field picking up the fallen grain. The continuously fed binding table allowed the binder to perform this task as the grain was brought to him or her. In 1880, a commercially viable mechanized binder was developed, eliminating the many hours which had been spent in binding.[96] By 1880, therefore, a single farmer could harvest wheat and bind it into sheaves by simply driving a horse-drawn reaper across the fields.

The threshing of wheat had also been mechanized by the mid-nineteenth century. Horses on treadmills were first used to drive the threshing machines, but after the Civil War steam engines gradually replaced the horse tread as the motive force.[97] These machines were capable of threshing thirty bushels of wheat per hour, compared to the seven bushels that could be threshed in an hour by a man swinging a flail, the traditional method.[98] The cost of a threshing machine, including the horse tread, was $230 in 1839, but dropped to $175 by 1851.[99]

Along with these various mechanical innovations in the production of wheat came a wide array of devices and implements which accelerated the accomplishment of all other agricultural tasks. Horse-drawn grass cutters for making hay were perfected in the 1850s, and horse-drawn rakes, tedders, and forks, all used for haying operations, were also available. Machines were also developed for harvesting, shelling, and crushing corn, pressing cheese, and so on. Without continuing this list or going into any further detail, there is little doubt that by the time the federal government began giving land to settlers in 1862, agricultural practices in the United States had been thoroughly transformed. There are three features of this transformation I want to stress.

First, the various farm tasks which had traditionally been performed by the farm family and hired hands had come to be treated as impositions on the use of time. Every task upon which time had to be spent became an obstacle to be overcome through technological ingenuity and animal power, the latter which would eventually be replaced by steam, and then gasoline, engines. In other words, farming had become convenient, in the modern sense of this word. The production of food as a means of satisfying a demand of the body was transformed into an array of limits to be overcome through technology.

This is not to claim that all American farming prior to the mechanization of the nineteenth century was subsistence farming, concerned only with the bodily demands of the household. But those farmers who first settled west of the Appalachian Mountains in the eighteenth and early nineteenth centuries—ahead of the surveys and land auctions—produced agricultural items primarily for household consumption.[100] In the course of the nineteenth century, however, as American space became increasingly ordered according to the U.S. land policies, farmers came to be less concerned with production for household consumption than with production for commerce. In the twentieth century, this shift has been carried to the point where farmers produce only cash crops and purchase their household foodstuffs from a grocery. But what I really want to stress here is not the subsistence/commercial distinction as much as the demand/limit shift.

Of course, in regard to limits, the situation of the nineteenth century farm household is not the same as that of the modern household described in Chapter 2. Overcoming temporal limits in the production of food for commerce is not the same as overcoming limits imposed by the preparation and consumption of food in the household. There is a difference between the nineteenth century farmer's purchase of a reaper, which speeded up production, and the modern household's purchase of a microwave oven and frozen foods, which accelerates consumption. But this leads to the second point I want to stress: the quickening of agricultural production in the nineteenth century was accomplished by turning the farmer, the traditional producer, into a consumer of technology. In Marxist terms, in the nineteenth century the production of food became a Department II enterprise, which produced commodities for mass consumption while consuming the means of production produced in Department I. By 1860, the production of farm implements and machinery (Department I agricultural production) had grown to be one of the top ten industries in the country.[101] Farmers had become not just commercial producers but commercial consumers as well.

Aglietta and Mandel both point out that, in regard to late or advanced capitalism, the integration of these two departments, along with the integration of individual consumers and Department II, depends to a large extent on the availability of credit. This holds true for the integration or subduction of independent farmers into the capitalist economy of the nineteenth century as well, and this is the third point I want to emphasize. The threat posed to capitalism by the vast amount of unsettled space, which might have allowed farmers to gain some independence from capitalist relations of production, was undermined largely by the extension of credit. First, credit was extended by both the government (1800–1820) and private lenders to the settlers who sought land upon which they could establish a homestead. Then, in order to pay back their land debt, the farmers had to buy machinery which would speed up production to the point where they could quickly bring all of their land into commercial use. And the purchase of agricultural machinery was, like the purchase of land, accomplished on credit.

Cyrus McCormick, who had developed the first commercially successful reapers, as well as a highly mechanized process to produce them, was also an innovator in marketing the new machines. He introduced the practice of selling farm machinery on credit. In 1856, two-thirds of his sales were made in this manner.[102] To keep pace with McCormick's sales, his competitors had to follow suit and offer credit to their customers. But even where credit was not available from the producers themselves, banks and other private lending agents were willing to loan the farmer the money needed to purchase machinery.[103]

This is hardly surprising, given the prices of the various machines which have already been mentioned. Reapers and seed drills were around $125 apiece in 1850, and a threshing machine was about $175. In comparison, the 160-acre parcel purchased by the settler in 1850, under the terms of the Pre-emption Act of 1841, would have cost $200. An investment in the new farm machinery was on the same scale as the investment in the land itself.

By the mid-nineteenth century, therefore, farming had been transformed not only into a primarily commercial endeavor, but into a commercial endeavor which required the consumption of expensive time- and labor-saving technology. There was no danger posed to capitalism in 1862, when the federal government began to give land to settlers free of charge. Even if these settlers were not burdened by the land debts that weighed upon their predecessors, their commercial success depended on their ability to produce crops as cheaply as those farmers who used the new machinery. If the settler wanted to sell even a small amount of the farm's total output, he or she had to be able to keep the cost of production down to a competitive level, and this meant employing the new technology.[104]

This discussion of nineteenth-century agricultural technology may seem irrelevant to an understanding of the techno-fetishism of the late twentieth century, since in technologically advanced countries relatively few people are engaged in agricultural activity. But that is the peculiar nature of agricultural technology; the more successful this technology was in freeing up the time involved in the production of food, the less visible and prevalent this technology became, as fewer people were required to spend their time in agricultural production. Nevertheless, the nineteenth-century advances in agricultural technology are important for understanding the role convenience plays in modern techno-fetishism. It was the success of agricultural technology in reducing the time that had to be spent in agricultural production which enabled technology to further establish itself in the modern household. It did this in two related ways.

First, the increased productivity provided by agricultural machinery allowed a greater proportion of laborers to become employed in other industrial enterprises, many of which produced items consumed by households. And many of the consumer items that emerged throughout the nineteenth century were conveniences. This leads to a second way in which the proliferation of farm machinery fostered further technological consumption—the consumption of various technological apparatuses which quickened and lightened the many tasks and chores of a large farm helped to groom or train individuals for the consumption of other time- and labor-saving devices which were not strictly agricultural. In other words, advances in agricultural technology, along with a land distribution system which fostered the consumption of that technology, promoted—but did not cause—the production and consumption of other forms of household technology.

I must stress here that the role which I ascribe to farm technology in the development of modernity's techno-fetishism is not a determinate one; there are no iron chains of causality or necessitation here. I am not claiming that the industrial production and consumption which have characterized the nineteenth and twentieth centuries were possible only because of technological breakthroughs in agriculture, nor am I saying that the blooming industrialism in nineteenth-century America required the mechanization of agriculture. I am claiming only that the mechanization of agriculture and nonagricultural industrialization under capitalism were complementary, and I refuse to reduce either one to a simple effect of the other. But most importantly, I want to add a third element into this blend—the problem of American space. The mechanization of agriculture actually began in England among the commercial farmers who owned the land that had been removed from common use. But it was in the United States that this mechanization became most fully developed. In England there was plenty of cheap agricultural labor among that class that had been "torn from the soil," as Marx put it. In the United States, however, there was plenty of space, at least in the first half of the nineteenth century, upon which settlers could produce for their own benefit. The mechanization of agriculture helped to undermine that threat while preparing a foundation for American techno-fetishism.

CHAPTER 5

Setting Bodies in Motion

The restrictive, regulatory land policies adopted by the U.S. government and the mechanization of agriculture are two closely related facets of capitalism's response to the problem of American space. A third facet of this response must be examined. I alluded to this dimension earlier when I discussed the commercialization of agriculture and pointed out that such commerce depended upon developments in transportation. But the point I want to make about transportation is not that commercial farming required, or was made possible only because of, an elaborate transportation network and rapid means of transportation which facilitated the movement of food and industrial products. In fact, it is not the movement of things that I want to primarily stress, but the movement of people. So far, I have presented the problem of American space as a threat posed by independent settlers who might have produced for themselves, not capital. But there is more to the problem of space than this alone. It is not only the settlement of space which presents a problem for capitalism; movement itself is also problematic.

Although Marx did not approach the problem of movement from the perspective of the spatial situation of the United States, he did recognize that the regulation of movement was a concern for capitalist development. In his examination of the "Bloody Legislation Against the Expropriated" (in the first volume of *Capital*), Marx does catalog some of the brutal British punishments meted out to vagabonds, those wanderers who had to resort to begging to stay alive. A brief examination of the British response to the problem of movement will help to set in relief the very different response of the United States to the problem of movement.

Although peasants were in a sense set in motion by the enclosures which occurred in England, the movement of these people was not in any sense free. These wandering beggars were beaten, mutilated, imprisoned, or forced into servitude if they stopped too long in any place.[1] But aside from laws against vagabondage, which tended to keep beggars in motion, there were also settlement laws, which were designed to keep the landless peasants from moving out of the parish to which they 'belonged.' Marx does not mention these settlement laws in his discussion of the expropriated, but Adam Smith does complain of them at some length in *The Wealth of Nations*.[2] Settlement laws prohibited any parish member from moving to another parish unless he had guaranteed employment or housing in that parish. If he did not meet these provisions, he could, within a forty-day period, be expelled from the parish to which he did not belong and returned to his original parish.[3]

Another interesting point about movement in England, a point that highlights the difference between the spatial situation of England and that of the United States, is that one of the prevalent forms of punishment for those who resisted enclosure was called "transportation." The usual punishment for the poaching of game from private, enclosed land, for example, was transportation. Transportation was also the punishment meted out to many of the agricultural laborers who revolted around 1830 and destroyed the threshing machines which were eliminating their jobs.[4] When a convict was punished by transportation, this meant that he was sent to a colony for a specified period of time, often for the duration of his life, during which he would labor in the service of a colonist. If the sentence was for less than life (seven years seems to have been the most common sentence), it was up to the convict to make his way back to England after serving his sentence.[5]

In the United States, on the other hand, during the same period in which rebellious English laborers were being "transported" as a punishment, the lower strata of American society were being induced to move. Transportation facilities were being established to make travel easier and faster for settlers heading west. But as I will try to show in this chapter, the transportation developments in the United States also had their disciplinary, regulatory aspect. To aid in presenting this interpretation of modern transportation, I will enlist the support of Paul Virilio, a writer who has paid attention to the way in which movement, especially mass movement, has been used and regulated by capital and the state.[6]

In broad terms, Virilio claims that modernity has been characterized by a shift in political and economic authority's use of the movement of the masses. Prior to the nineteenth century, writes Virilio:

> Society was founded on the brake. Means of furthering speed were very scant.... In general, up until the nineteenth century, there was no production of speed. They could produce brakes by means of ramparts, the law, rules, interdictions, etc. They could brake using all kinds of obstacles.[7]

Then, however, occurred the revolution, characterized in so many ways by so many writers, which ushered in modernity. From Virilio's perspective, the significance of this revolution was not that industrialization introduced mass production nor that liberal democracy permitted mass politics, but that speed production became possible. "And so they can pass from the age of brakes to the age of the accelerator," writes Virilio. "In other words, power will be invested in acceleration itself."[8]

Given his unique perspective, Virilio describes the revolution which opened up the modern period not as an industrial or democratic revolution, but as a "dromocratic" one. As he puts it, somewhat excessively: "In fact, there was no 'industrial revolution,' but only a 'dromocratic revolution'; there is no democracy, only dromocracy; there is no strategy, only dromology."[9] The point of Virilio's substitutions, which are based on the Greek *dromos*, meaning running (or race) course, is to emphasize the revolution in movement which characterizes modernity.

Although I am not going to defend Virilio's hyperbolic claim that there was no industrial or democratic revolution, I do want to develop his claim about dromocracy. The affinity I have with Virilio centers on his insight that the freedom to move, an achievement of modernity, has become an obligation to move.[10] And although Virilio focuses primarily on military developments in transportation,[11] not on the developments in the realm of consumption, he does mention the importance of the American automobile for the dromocratic revolution.

Like Aglietta, Virilio recognizes that the mass production of the automobile was capable of transforming consumption practices.[12] But when Virilio does consider the consumption of movement, he, also like Aglietta, neglects those developments in transportation which preceded the mass production and consumption of the automobile. These developments in transportation technology, along with the concurrent developments in farm technology, helped to undermine the threat of American space and prepare the way for techno-fetishism and in particular the consumption of technologies of speed. The brief description of American transportation developments I will now offer should make this last claim clear while extending Virilio's unique perspective a little further into the realm of consumption.

The first westward movement of American settlers away from the Atlantic coast was accomplished by following the example and routes of the indigenous population. Narrow footpaths, called traces, connected the various navigable streams and lakes. Settlers walked along these traces, and used pack animals (oxen, mules, or horses) to carry their household supplies.[13] When they reached a waterway that led toward their destination they would purchase, rent, or construct a canoe made from a large log, into which they would pack all their possessions.[14] In those areas which could be reached by traveling up the Connecticut and Hudson Rivers, sailing vessels were used to bring the settlers' supplies close to their new home, but the settlers themselves and their livestock followed the traces through the forest.[15] It was in this manner that Connecticut, Massachusetts, New York, and Pennsylvania were settled in the seventeenth and most of the eighteenth centuries.

Toward the end of the eighteenth century, after the Revolution, the movement of settlers into New York and Pennsylvania, and beyond to the public domain in the Ohio River Valley, began to increase.[16] As these settlers established themselves in the new area, they also widened and cleared the land routes over which the indigenous population had earlier traveled, turning the traces into pack trails.[17] These pack trails became arteries for commerce as well as the westward movement of settlers. But the movement along these trails was at first slow and unpredictable. Fallen trees, washouts, and floods were obstacles which could emerge unexpectedly on any journey along these trails.

While these routes were frequently obstructed by the effects of the weather, they were, on the other hand, free of any effective political obstructions, such as the settlement laws of eighteenth-century England. Although it is true that George III did prohibit American colonists from moving beyond the Appalachian Mountains, and the U.S. government prohibited movement into unsurveyed areas of the public domain, these attempts at restricting westward movement were ineffectual due to the vast and unorganized space of America. Trappers, miners, squatters, and legitimate settlers were able, notwithstanding natural obstacles, to move freely along those trails which had been etched into the land over the years. But toward the end of the eighteenth century, this unregulated westward movement began to be set in order, just as the land itself was being set in order through rectilinear surveys. Regulation, however, did not take the form of a prohibition or restriction of movement, but an acceleration. By making travel along these routes more convenient—that is, easier and faster—such travel was also brought under control and integrated into the established order.

The manner in which this regulation of movement was initiated was the granting of turnpike charters by state governments to private companies.[18] These companies would improve existing routes by grading and draining them and replacing their dirt surface with gravel, which was sometimes placed on a firm bed of larger stones, a process known as "macadamization."[19] These roads were more resistant to the ravages of rain than were dirt roads. In particularly wet areas, corduroy roads were built by laying logs tightly together, perpendicular to the flow of traffic. Charters were also granted for the construction and operation of bridges and ferries.[20] As a result of these improvements, the various pack trails leading west from New England into Pennsylvania and New York, as well as those connecting the commercial centers of these states, were transformed into roads over which wheeled vehicles, such as stage coaches and wagons, could travel.

One of the first and most successful of these turnpikes was the one connecting Lancaster and Philadelphia in Pennsylvania, which was completed in 1794.[21] It was over this route that the Conestoga wagons established their reputation as the most efficient means available for transporting goods overland. These large, sturdy, wide-wheeled wagons were developed around Lancaster in the middle of the eighteenth century and were essential to the settlement of the public domain. These wagons, or facsimiles, were the ones used in the wagon trains which brought settlers into the West and were the prevalent means of land transportation until the middle of the nineteenth century.[22]

It was not until the turn of the century, however, that these horse-drawn wagons were able to travel beyond the Appalachian Mountains. The Wilderness Road (originally known as Boone's Trace, after Daniel Boone, who first marked its course in 1774–1775) was the primary land route into the Northwest Territory during the eighteenth century. But this road did not become suitable for wagon travel until 1795, when the Kentucky legislature passed an act requiring that improvements be made in the trail.[23] Even then, the Wilderness Road was not constructed out of gravel and stone, but remained a dirt road.[24]

The project of establishing a macadamized road through the Appalachian Mountains was undertaken early in the nineteenth century by the federal government. Construction of the National (or Cumberland) Road began in 1808 and was completed from Cumberland, Maryland to Wheeling, Virginia on the Ohio River in 1818.[25] The road was eventually extended, as originally planned, through the state of Ohio and into what would become the state of Illinois.[26] Ultimately, the federal government relinquished its interest in the road, and turned it over to the control of the states through which it passed.[27] Nevertheless, it remained, until the middle of the century, the main artery through which western settlers, as well as their commercial products, had to move.[28]

Alongside this massive public transportation project, many private turnpikes were constructed in Ohio, under charters issued by the state. The state government frequently invested in these projects,[29] alongside private stockholders of the companies. And the federal government still was involved in financing these improvements, although indirectly, through its land policy. Since the admission of Ohio to the union in 1803, three percent of all sales of public lands went to the states in which the sales occurred, and this money was earmarked for development of roads (and eventually canals).[30]

This first step in improving transportation, or facilitating movement, in the territory of the United States, appears to lie somewhere between Virilio's characterization of the modern and premodern eras. The new techniques of road construction greatly increased the speed of movement throughout the frontier by allowing wheeled vehicles to cross terrain that had been previously suited only to pack animals. Whereas pack trains could cover twenty-five miles in a day,[31] stage coaches could travel much farther in the same period, moving at a rate of six to eight miles per hour.[32]

But this acceleration was accomplished with the simultaneous obstruction of movement, a characteristic of the premodern period. On the National Road, "at average distances of 15 miles toll-houses were erected and 'strong iron gates hung to massive iron posts were established to enforce the payment of toll in cases of necessity.'"[33] On some of the private turnpikes, tollgates were established every four or five miles.[34] It is as though the acceleration provided by the stone and gravel roads was both an advantage and a threat to the existing order. The westward movement of settlers still had to be restricted, and the settlers' connection with the capitalist economy reinforced periodically by the payment of tolls throughout their journey. (Of course, the largest portion of the tolls collected was provided by commercial traffic, but my concern here lies primarily with the way in which transportation improvements helped to control and regulate the movement of the settlers themselves. Once the settlers had established homesteads, their commercial activity, which was intensified by land policies, was indeed regulated by the turnpikes; but in this discussion of transportation, I want to emphasize how the movement of people was regulated.)

Along with the above improvements in land travel, various developments in water travel also occurred around the turn of the century. In the late eighteenth century, the usual method of traveling by water into America's unsettled land was by flatboats. When settlers finally reached the Ohio River, either by wagon on the Wilderness Road, or by pack-train along one of the several routes through the Appalachian Mountains, they usually built their own flatboats, in which they would continue their journeys.[35] These large boats, which ranged from twenty to sixty feet in length and ten to twenty feet in width, carried families of settlers and all their household possessions, including livestock, downstream with the current. The settlers often spent weeks on these floating barnyards.[36] Since these boats were incapable of traveling upstream, they were usually dismantled once their destination on the river was reached.[37]

For traveling upstream, a different type of boat was developed, one which had a V-shaped hull, at the point of which was attached a wooden keel which ran the length of the boat. The keel protected the boat in case it ran aground and also allowed the boat to remain stable when headed upstream, as it displaced the current. These keelboats were propelled upstream by several different methods, all of which relied on manpower.

In weak currents the boat could be rowed, while in stronger currents poles were often used to propel the boat upstream. Another method used in strong currents was to attach a long rope to the boat, and then walk along the bank of the river, pulling the boat along.[38] In any of these cases, upstream travel was very arduous and time-consuming. It took one month to float downstream from Pittsburgh to New Orleans, but it took four months, and a crew of four to twelve, to return by keelboat.[39] Consequently, that movement of settlers into the frontier region which was accomplished by water was, during the eighteenth century, primarily headed downstream.

This situation changed with the introduction of steamboats onto the western rivers. In 1811, the *New Orleans*, a steamboat built by Livingston and Fulton's Company, set off from Pittsburgh to New Orleans. Upon reaching Louisville, where the flow of the Ohio was broken by a series of falls, the *New Orleans* had to wait for the water level to rise. During this wait, the boat traveled back upstream to Cincinnati, demonstrating its ability to move quickly against the current of the Ohio River.[40] (Steamboats had already been operating on the rivers of the East, such as the Hudson, but their ability to handle the larger, more treacherous rivers of the West had been in doubt.[41]) In 1815, another boat owned by the Livingston and Fulton company, the *Enterprise*, steamed its way from New Orleans to Louisville in twenty-five days, but under unusually favorable conditions.[42] The same trip by keelboat would have taken around three months.[43] In 1817,

Henry Shreve's *Washington* completed a round trip from Louisville to New Orleans in forty-one days under normal conditions, demonstrating the feasibility of regular steamboat travel up the Mississippi and Ohio Rivers into the Northwest Territory.[44]

After this point, steam navigation on these two rivers greatly expanded. In 1817, there were seventeen steamboats on the Ohio and Mississippi Rivers. By 1820, there were sixty-nine, and by 1855, seven hundred twenty-seven steamboats were plying the waters of these two rivers and all of their major tributaries.[45] And the speed of these boats increased as their numbers did. In the round trip made by the *Washington* in 1817, twenty-five of the forty-one days were spent going upstream from New Orleans to Louisville. By 1820, that upstream trip could be made in ten or eleven days, and by 1853, it was possible in less than five days.[46] The steamboat, more than the turnpike, is a clear example of the speed and acceleration which Virilio stresses in his interpretation of modernity. These boats attained speeds of ten miles per hour on the western rivers, and even faster speeds in the East.[47]

While these boats were primarily used for transporting commercial goods up and down the river, they were nevertheless crucial to the movement of settlers into the Mississippi Valley. By the 1850s, the boats were large enough to carry three to four hundred deck passengers, many of whom were immigrants moving into the public lands of the Mississippi region.[48] But even though the development of the steamboat provides a good example of the acceleration of the movement of settlers, there was a certain weakness to the steamboat in regard to its ability to regulate that movement.

Unlike the turnpikes, which could thoroughly regulate movement along their routes through the use of tollgates, steamboats were unable to completely control the movement along the rivers. Along the turnpikes, even those travelers who provided their own means of conveyance and avoided using the stagecoach and freight companies to move themselves and their belongings still could not avoid paying the tolls.[49] But on the rivers, settlers could build or purchase a flatboat or keelboat and move themselves independently from (although more slowly than) the steamboat companies. In fact, the number of flatboats on the Mississippi and Ohio Rivers reached its peak during the late 1840s, when steamboats were spreading out into all the tributaries of the Midwest.[50] For despite the development of steamboats, the rivers remained uncontrollable, natural routes for transportation. Their banks offered many points from which flatboats and keelboats could be launched and landed, and it was impossible to charge these vessels for their movement along the river. However, the other major development in nineteenth-century water transportation—the construction of canals—overcame this and other problems posed by unruly rivers.

Between 1817 and 1845, the construction of canals was carried on at a frantic pace in the United States. In 1816 there were barely one hundred miles of canals in the country, but by 1840 there were more than thirty-three hundred miles.[51] Many of the canals built during this period (as well as the pre-1817 period) were designed to overcome a particular obstacle in a river, usually a falls.[52] In fact, in 1828 one such canal was built around the falls in the Ohio River at Louisville, making it possible for steamboats to travel from the Mississippi to the upper reaches of the Ohio River.[53] In these instances, the canals overcame the hazards posed by a particularly steep pitch in the course of the river by establishing an alternative waterway that contained a series of locks. These locks would gradually float boats up or down the dangerous incline, eliminating one of the chief impediments to river travel.

Other canals, however, were built not to overcome obstacles in rivers, but to create waterways where none had been before. Such canals were often built to connect two water routes, such as an inland and a tidewater river. These canals were actually artificial rivers, constructed where a river would have been beneficial to commerce. Canals of this sort were constructed around eastern commercial centers in the first four decades of the nineteenth century, but they were not widely used to transport travelers.[54] In 1817, however, the state of New York began construction on the Erie Canal which, upon completion in 1824, connected Lake Erie with the Hudson River. Although the Erie Canal was an important commercial link between the Great Lakes and the Atlantic Ocean, via the Hudson River, it was also the first canal widely used for the transportation of people.[55] The canal carried settlers not only to the largely settled territory along its 364-mile course across the state of New York,[56] but it also served as a major transportation route for settlers heading for the Ohio River Valley.[57]

The success of the Erie Canal caused Pennsylvania to undertake construction of a competing canal route from the East to the public domain of the West. This canal was even longer than the Erie, at 395 miles,[58] and it also faced an additional obstacle. The route of the canal from the Susquehanna to the Ohio River ran directly across the Allegheny Mountains. In order to cross these mountains, a series of inclined planes was constructed on each side of the mountain range, and tracks were laid on these planes. Stationary engines pulled passenger and freight cars up one side and lowered them down the other, and eventually these engines moved sections of specially built canal boats over the mountains.[59] The Pennsylvania Canal was begun in 1826 and completed in 1834, and cost over ten million dollars. However, it was never as successful as the Erie Canal, partly due to the portage railroad, which was a bottleneck in the canal route, and partly because of railroad competition.[60] Nevertheless, the Pennsylvania Canal (or Main Line Canal, as it was often called) stands as testimony to the faith Pennsylvania had in the benefits that would be provided by a canal route into the public domain.

Although travel through the canal was not as fast as travel by river steamboat (canal boats averaged speeds of three to four miles per hour[61]), the canals, like the steamboat, did eliminate much of the toil and trouble associated with upstream travel. With their system of locks, the canals practically eliminated the current one would normally face when traveling from a lower to a higher point by water. And in the canals there was no need to exert human power against even the minimized current; the boats were pulled along by horses or mules which walked along the towpaths lining each side of the canal.

Another feature of the canals, one which may have helped convince New York and Pennsylvania to expend so much on their construction, was that they were, in a sense, like turnpikes of water. The banks of the canals, unlike those of the rivers, did not provide easy, unregulated access to the water. And the many locks on the canals (84 on the Erie, 174 on the Pennsylvania) were, like the tollgates on turnpikes, stations at which payment could be exacted from travelers moving themselves in their own boats, independently from the private freight and passenger businesses which developed on the canals. For those who booked passage with one of the canal businesses, the tolls were included in the fares which were charged.

As I have already mentioned, the Erie and Pennsylvania Canals were financed by the states through which they ran. It will be recalled that it was during this period—the 1820s—that the federal government began to curtail its direct involvement in the settlement of the public domain. In 1820 the federal credit system for land purchases was abolished (see Chapter 4, p. 77), and in 1829 the federal government began to relinquish control of the National Road to the various states through which it passed (see p. 96 of this chapter). But in the construction of canals the federal government developed a new method for promoting the settlement of American space, a method that would become crucial in the construction of railroads.

In order to encourage the construction of canals in the territory the federal government had put up for sale, land grants were given to the young states which had been established there. Ohio, Illinois, and Indiana—the states of the Northwest Territory—all received from the federal government not only the right of ways upon which to construct certain canals, but also half of the land for five miles on each side of these canals. The other half of the land was retained by the federal government, the land being divided into odd- and even-numbered sections (640-acre lots) and distributed on an even/odd basis between the state and federal governments. Odd-numbered sections belonged to the states, even-numbered ones to the federal government. The states could then sell their sections and use the revenue to finance the construction of the canals, or they could grant this land to private companies, who could then sell the land for the same purpose, and for a profit as well.[62]

The period of canal construction was rather short-lived, however, because developments in rail transportation around the middle of the nineteenth century overshadowed all other forms of transportation. Railroad vehicles, when driven by steam engines (not horses, as they were originally[63]), offered the possibility of very rapid transportation. The passenger trains of 1850 were able to travel around twenty-five miles per hour.[64] Furthermore, railroads, like canals and turnpikes, were not restricted to any natural course, such as a river. But unlike the canals and turnpikes, which allowed independently owned boats and wagons to travel their routes (for a price, of course), the railroad, with its steam locomotives, precluded such independent movement.[65] To travel by rail meant to travel on the terms, over the route, and at the rate of speed established by the railroad companies. More than any of the other forms of transportation examined thus far, the railroad exemplifies the acceleration and speed which characterize modernity as well as the regulatory dimension of this rapid movement.

It is also with the railroad that the close relation among developments in transportation technology, the land policies of the United States, and the reification of American capitalism becomes undeniably apparent. As was the case with the federal land grants for canals, the first land grants for railroad construction were given to states, not entrepreneurs. But the scope of the railroad land grants was, from the start, beyond that of the canal grants.

In 1850, Congress granted to the states of Illinois, Mississippi, and Alabama, for the construction of the Illinois Central Railroad, the odd-numbered sections of the land six miles on each side of the route. This amounted to a grant of more than two and a half million acres.[66] Throughout the 1850s, similar grants were given to various states, totaling around eighteen million acres.[67] But these land grants, many of which were given for the construction of local railroads, were only a precursor to the much larger and more significant grants given for the construction of transcontinental lines.

The transcontinental land grants (the first occurred in 1861, the year before the passage of the Homestead Act) were not only longer than the grants of the 1850s, but they were wider as well. The initial grant of this sort was given for the construction of the Union Pacific Railroad and was comprised of the odd-numbered sections of a twenty-mile-wide tract of land along the route. This single grant amounted to around twelve million acres.[68] The grant made to the Northern Pacific, the last of the transcontinental grants, was almost four times the size of the Union Pacific grant. For the part of its route that traversed existing states, the Northern Pacific was granted half of the sections of a forty-mile tract, and in the territories the tract was doubled to eighty miles. This grant amounted to over forty-seven million acres.[69]

Aside from size, there was another important difference between the grants of the 1850s and the transcontinental land grants; the latter were given directly to the railroad companies, not to the states through which the tracks would pass. All told, about ninety million acres of land were granted to such private corporations by the federal government.[70] The railroad companies, therefore, were not just in the transportation business; they were also the largest land companies in the nation and sold homesteads to those settlers they moved across the nation. The railroads controlled to a great extent both the movement across and the settlement of American space.

In fact, the transcontinental railroad companies actively sought colonists for the settlement of the western territory and had large forces of agents in Europe. These agents would induce members of the peasant and lower-middle classes of Europe to immigrate, but not for the purpose of swelling the labor force of the eastern United States (as one would expect according to Wakefield's plan for systematic colonization), but rather to become western farmers.[71] Not only did these agents spread the word about the opportunities available in the midwestern United States; they often offered reduced train fares to settlers or allowed them to deduct the price of their fare from the price of any land they purchased from the railroad company.[72]

And once the settler reached the western territory, the railroad companies made it relatively easy to acquire land within the grant area. Forty-acre lots were often sold to settlers and at prices not extremely higher than the $2.50-per-acre price at which government land in the grant area could be pre-empted, prior to the passage of the Homestead Act.[73] The railroad companies also extended lenient credit terms to the settlers, and some companies required only interest (usually around 6 percent) and not principal to be paid for the first years after settlement.[74]

The lenient land sale policies of the railroad companies can be explained in part by the passage of the Homestead Act. After passage of the act, free land was available outside the grant area as well as within, and this limited the price the railroad companies could demand for their land. But it must be recalled that much of the best agricultural land had already been claimed by 1862 (see Chapter 4, p. 81). Furthermore, agriculture had been mechanized and commercialized by this time, and farmers could not afford to settle on land twenty or forty miles from the railroads. So the effect of the Homestead Act on land prices should not be overemphasized. Other factors help to explain the eagerness with which the railroads disposed of their land.

One of these factors has to do with the hermetic nature of rail travel itself. As I have mentioned, railroads not only moved people and things more quickly than any other existing form of transportation, but they moved them in a highly regulated manner. Unlike the other forms of travel which have been examined, railroads allowed no independent movement. Once the settlers were established, any commercial activity they undertook was completely dependent on the railroads. In order to sell their produce, farmers not only had to save time in the fields through the use of farm machinery, but they had to get their goods to market as quickly as possible through the use of trains. The lenient land terms offered by the railroad companies can be interpreted, therefore, as a railroad company decision to sell its land quickly and cheaply, rather than hold out for a higher price, so as to quickly begin reaping the long-term profits which it could extract through its transportation monopoly.

Of course, the railroad monopolies were challenged in the last decades of the nineteenth century by the concerted efforts of farmers (e.g., the Patrons of Husbandry, or the Grange, and the Populist Movement). But even when the farmers were successful in resisting these monopolies (e.g., in getting states to regulate the railroad's management of grain elevators[75]), such achievements, just like the Homestead Act, can be viewed not so much as victories of the farmers over capital as the death throes of the agricultural challenge to capital. These late-nineteenth-century agricultural movements would not be as successful as their predecessors, the claim clubs, were in their resistance to land speculators. In the time that had lapsed a certain amount of order had been imposed upon American space, and an important dimension of that order, I am arguing, was the regulation of movement which was accomplished through technological developments in transportation. Through the consumption of this transportation technology, homesteaders took their place in this order. Although it may appear that I have overplayed the regulatory dimension of acceleration or speed, a brief summary which juxtaposes developments in land policy and transportation technology will lend some indirect support to my claims.

When the U.S. government first acquired its vast land holdings, squatters were moving into these territories by means of pack trains and flatboats and were clearing and cultivating small homesteads in the most desirable areas. The government's response to this unregulated westward movement and settlement was the adoption of restrictive land policies. Military force was used (although infrequently) to evict squatters from government land, and the homesteads were destroyed. Settlement could only occur, according to the land policy of the government, after the land had been officially surveyed into square townships and sections, and the settler purchased at least a full section at auction. The restrictive nature of this policy becomes obvious when one considers that in 1785, when this policy was first adopted, none of the farm machinery discussed in the previous chapter was available to the settler. The 640-acre minimum was eight times the minimum purchase of the 1820s, when durable plows with replaceable iron parts were speeding up the process of clearing the land. The original minimum of 640 acres, along with the minimum price requirement, prevented widespread legal settlement by small homesteaders.

But throughout the first two decades of the nineteenth century, the federal government relaxed the restrictions on settlement which were imposed by the Ordinance of 1785. The minimum purchase size was gradually reduced throughout this period, until it reached the low of eighty acres in 1820, and the government offered easy credit terms to settlers during these years. It was also during this period that the Wilderness Road and the National Road had made it possible for wagons and stage coaches to pass through the Appalachian Mountains, and steamboats were well established on the Mississippi and Ohio Rivers by 1820.

The decade of the 1820s saw the opening of the Erie and Pennsylvania Canals, and in 1830 the first American passenger railroad was opened (see note 63 of this chapter). In regard to land policy, 1830 also marks several important developments. With the Pre-emption Acts of the 1830s, the federal government reversed its longstanding position on squatters' rights and allowed many squatters to buy the land on which they had settled. The government also reversed its practice of establishing minimum lot sizes. Although these minimums had been decreasing throughout the first decades of the nineteenth century, with the Pre-emption Acts the government began establishing maximum size limits. A settler no longer had to buy *at least* a specified number of acres; he or she could now buy *no more* than 160 acres (see Chapter 4, p. 79). I read this shift as another indication that the threat of American space had largely been overcome. Farmers were by this time so caught up in the existing order that it was safe to promote their settlement on public land in large numbers. The maximum size limit would enable even more people to settle within the order that was being established in America.

1861–1862 is the other important period at which changes occurred in land policy and transportation development. It was in 1861 that the first transcontinental land grant was given to the Union Pacific Railroad company by the federal government, and 1862 is the date when the government finally began to give land to homesteaders. I read these events as a late phase of the confrontation between capital and American space. Once movement and settlement were thoroughly regulated by the railroad monopolies, with their extensive land holdings, land policy no longer had to serve its original restrictive function. Since settlers had been transformed from independent, hence dangerous, producers, to consumers of technological conveniences, the threat of American space was overcome and land could safely be given to those who labored upon it.

This reading is further supported, I believe, by the manner in which the Homestead Act was applied in the railroad grant areas. In these areas, where the railroads held virtually complete sway, the government tried to increase the number of settlers even more than it did in other areas of the public domain. The Homestead Act was applied to the government-owned sections in these grant areas, so settlers could get their land for free, but the maximum size limit was reduced to eighty acres,[76] allowing even more people to settle in these 'safe' zones. And to ensure that these areas were indeed rendered safe by the railroad companies, the government usually suspended settlement in any area that was under consideration as a route for the tracks, as earlier it had prohibited settlement in unsurveyed areas. Vast areas were withheld from homesteading until the railroads had set things in order.[77] After the railroad companies had decided on the location of the tracks, not to mention the townsites, junctions, and so on, it was safe to bring in settlers. When squatters interfered with the establishment of railroads in the grant areas, the government quickly reverted to its original, direct approach to the squatter problem: it used military force to eliminate this threat to order.[78]

What I have tried to indicate in the last two chapters is one dimension of a genealogy of convenience or, if one prefers, a genealogy of modern consumption. Although this dimension is concerned with material conditions, it is one that is overlooked by Marxists, even those Marxists who have begun to look at consumption patterns. It is a dimension which is clearly related to the reenforcement of capitalist relations of production, but it does not reduce to those relations of production. Rather, the consumption of farm and transportation technology, which blossomed in nineteenth-century America, resulted in part from capitalism's coming to grips with the danger or threat or American space.

One outcome of this confrontation between capitalism and open space was that agricultural activity became a limit which the body imposed on the use of time, a limit which was to be overcome through the consumption of technology. Through the land policies I examined in the preceding chapter, in conjunction with developments in agricultural machinery, farming was transformed from a largely self-sufficient activity into a primarily commercial endeavor. The production of food came to require the consumption of ever-increasing amounts of time- and labor-saving technology, or convenience. And all this had largely occurred by 1860, long before the period of over-capitalization which Mandel identifies as the condition which permitted/demanded the industrialization of agriculture.

On another front, the movement across American space was also brought under control in the nineteenth century. Old, slow travel routes and methods such as pack trails and flatboats, were supplanted by faster and more tightly controlled means of movement. American space was transformed from a possibility for freedom, for escape, into distance, another limit of the body to be overcome through the consumption of technology. And as I mentioned earlier, when the need to overcome distance is combined with the need to save time from bodily necessity, an endless series of limits develops. New technologies are required to move men and things not only across all distances, but to do so at an ever-faster pace. Thus, as the railroad replaced turnpikes and canals, it was in turn replaced by the airplane, which was then replaced by the jet. Jets are now being developed which promise to lessen even further the time spent traveling from continent to continent. And soon, no doubt, there will be passenger shuttles to nonterrestrial space, the new frontier.

However, the automobile, which is rightly emphasized by Aglietta as a hallmark of modern consumption (see Chapter 3, p. 58), does not at first glance fit within the perspective that I am developing here. It is a means of conveyance which allows travelers to overcome distance quickly, especially in local travel but unlike the train or the airplane, it does not seem to be amenable to close regulation. In an automobile, one can determine (to some extent) when and at what speed one will move. And one can (again to some extent) escape in an automobile (but a high-speed chase is likely to ensue). At the very least, the automobile can certainly be used for more than travel to and from the workplace. But there is more to the mobile existence which the automobile has provided or, one might say, demanded.

Ruth Schwartz Cowan has argued that the automobile, like so many other household conveniences, has not really liberated the time of women who work in the home, but has burdened them with an array of delivery and pick-up tasks which had been performed by men of the house or by delivery services.[79] And Ivan Illich has argued that "the product of the transportation industry is the *habitual passenger*,"[80] and he describes this product with a sense of loss and impotence rather than freedom. According to Illich, the habitual passenger is "addicted to being carried along" and "has lost control over the physical, social, and psychic powers that reside in man's feet.... He has become impotent to establish his domain, mark it with his imprint, and assert his sovereignty over it.... Left on his own, he feels immobile."[81]

The points I want to stress about the transportation possibilities introduced with the automobile are somewhat different from Cowan's and Illich's, however. I agree that the transportation system which developed around the automobile set people in motion and required them to move. But I also want to emphasize that this form of transportation technology binds people to the economic structures of modernity in a manner similar to that which bound the nineteenth-century homesteader.

Whatever freedom is provided by the privately owned automobile (and there is some), it is compensated for by the financial obligations one incurs through the consumption of the automobile. Cars are one of the most expensive consumer items purchased by households, and they are usually bought on credit. In this sense, then, the automobile marks the convergence of the two regulatory trends which have been identified in this genealogy. The automobile, like the other forms of transportation I have examined, accelerates the movement of people; and like farm machinery, the automobile binds the consumer with a debt that must be paid over time.

Along with strapping the habitual passenger with a credit burden, the automobile also introduces a vast new array of possibilities for the consumption of convenience. Fast-food restaurants, supermarkets, and drive-through windows of all sorts, from liquor stores to banks, have emerged in the culture of the automobile. For these reasons, the automobile is, at least for the present, probably the best symbol of the modern culture of convenience, where speed in overcoming the various limits of the body is a primary value.

The status of the automobile as the hallmark of modernity is also attested to by the fact that the acceptance of the automobile by consumers was itself a very speedy affair. Whereas other forms of household technology often took several decades to gain widespread acceptance, the automobile very quickly became an item of mass consumption. At the turn of the century, there were several car manufacturers producing expensive models, but by 1930, there were twenty-six million registered automobiles for thirty million households.[82] By the time Henry Ford was introducing his Model T (and refining the assembly line production system which bears his name), American consumers had largely been sold on the value of convenience and the ideal of speed.

At this point, the techno-fetishist will probably have had enough of this heretical babbling—the car as a restriction or burden!—and should be ready to put a halt to this line of thought. The perfect roadblock would seem to be available—or rather, a rotary, which can turn this argument back on itself. One might ask at this point whether this questionable interpretation of agricultural and transportation technology has not ended up treating the consumer in just the manner of which I complained in my treatment of current Marxist analyses of consumption? Indeed, is not my treatment of consumers even closer to the rigid Marxism of structuralists than the treatment of consumers offered by the several Marxists I discussed in Chapter 3? I make it sound as though the settlers on the American frontier were simply herded like cattle onto the various forms of transportation and fed farm equipment as if it were fodder. I appear to be oblivious to the possibility that the settlers, not to mention contemporary consumers of automobiles and other conveniences, may have actually wanted or needed this technology. With my argument, as with the Marxist argument, it is still capital which ultimately prevails and appears to thoroughly determine consumption and the values which underlie it.

My response to all of this, however, would be to point out that the last two chapters have presented one dimension of a genealogy of convenience. The resolution of the threat which open space posed to U.S. capitalism helps to explain, in part, the techno-fetishism of modernity. Contrary to the possible critique of my perspective mentioned above, however, I do acknowledge that there was indeed a need among individual consumers for the various technological developments which I have so far examined. This need, of course, is the need for convenience. My claim is not that American capitalism created this need in the nineteenth century or created consumers who valued convenience. My point is that American capitalism played upon this need for, or value of, convenience, and was thereby able to undermine the threat posed by unsettled land.

Of course, I am not claiming that this value of convenience is a uniquely American one; it is a value which characterizes modernity. But the spatial situation of the United States in the eighteenth and nineteenth centuries provided the conditions in which this value was able to flourish and helps to explain why modern consumption standards have developed principally in the United States. But these last two chapters have not shed any light on the proliferation of the value of convenience after the threat of American space was overcome or beyond the borders of the United States. In the next chapter, I will try to indicate the broader foundation of the value of convenience.

CHAPTER 6

Weber, Protestantism, and Consumption

THE VALUE OF WEBER'S ARGUMENT

The material considerations of the last two chapters were offered not simply as an argument against certain recent Marxist thought on consumption. They were offered primarily because of the influence that those particular material conditions had on the development—or rather, deployment—of the value of convenience. The spatial conditions of the United States in the nineteenth century had an exacerbating or accelerating influence on another development which had been taking place in other Western countries as well as the United States. What I have in mind here is not the development of capitalist relations of production, which Marxists have done so much to illuminate, but the devolution of Christianity. And if Marxists have tended to overlook the significance of American space in the development of late capitalism's "social norm of consumption," they have remained largely oblivious to the possibility that religious ideas, and the problems or situations to which those ideas are a response, can have any significant impact on the development of capitalism in modernity. Although the several Marxists I discussed in Chapter 3 have moved to varying degrees beyond the simple economic determinism of a more structuralist Marxism, they do not seem to have gotten much farther than Marx himself in recognizing any effective role of (not for) religion in the development of Western capitalism.

At this point, I am going to pick up a theme that was raised before I considered Marxist thought on consumption, a theme that emerged in my criticism of Hannah Arendt. But to recall this theme, which concerns Christianity and the body, it is best not to return to my argument with Arendt's interpretation of Christianity as a form of reverence for life. For Arendt sees such reverence not only in Christianity but in Marxism as well. And although I disagree with both of these claims, criticizing them will not get me very far in linking the materialism of Marxism with developments in, or transformations of, religious ideas. To begin again on this line of thought concerning Christianity, the best route is Max Weber's thinking on Protestantism and capitalism. For Weber's thoughts on Christianity and modernity are much closer than Arendt's are to the argument I want to make.

Weber's *The Protestant Ethic and the Spirit of Capitalism* has generated an exorbitant amount of controversy since its initial publication as a two-part article in 1904–1905.[1] And although I have not been shying away from controversy up to this point, perhaps I should offer some explanation or apology for my choice of such a battered text. My reading of Weber's text emphasizes the limitations he imposes on his argument. I pay close attention when he claims that he has "no intention whatever of maintaining such a foolish and doctrinaire thesis as that the spirit of capitalism...could only have arisen as the result of certain effects of the Reformation, or even that capitalism as an economic system is a creation of the Reformation."[2] And I count myself as part of the group when he claims that "we are merely attempting to clarify the part which religious forces have played in forming the developing web of our specifically worldly modern culture, in the complex interaction of innumerable different historical factors."[3]

I have to withdraw, however, when Weber, in a footnote toward the end of the text, abandons his earlier restraint and says of religious ideas "that they are in themselves, that is beyond doubt, the most powerful plastic elements of national character, and contain a law of development and a compelling force entirely their own."[4] I have no qualms with excessive argumentation (see the justification of my own excesses in Chapter 1, pp. 10–11), but when it becomes dogmatic instead of rhetorical, as it does in this last quote from Weber, I have to balk. So in the use to which I put Weber's argument in this text, I will be treating it as an argument of limited claims. Weber offers *a* perspective on modernity which emphasizes the role of religious ideas, and my aim in this and the following chapter is to examine and expand that perspective.

Weber's task in *The Protestant Ethic*, as I read it, is not just to identify the influence of the Reformation on the development of the modern, rational spirit of capitalism.[5] There is little doubt that the main claim of his argument is that Protestantism, especially Calvinist Puritanism, spawned a new breed of entrepreneurs who threw themselves into business life without the slightest religious compunction. What the Reformation bequeathed to its secular successor, claims Weber, was "above all an amazingly good, we may even say a pharisaically good, conscience in the acquisition of money."[6] But Weber also recognized the legacy of the Reformation among those who labored for the new breed of entrepreneur and were concerned less with the accumulation of wealth than with satisfying the needs of their households. The Protestant idea of "labour as a calling became as characteristic of the modern worker as the corresponding attitude toward acquisition of the business man."[7]

So what Weber was trying to get at in his controversial text was not just the rise of the modern entrepreneurial spirit. More broadly, he was concerned with the emergence of individuals of a sort amenable to the rationally organized capitalism of modernity, which Weber famously characterized as an "iron cage."[8] Weber describes his task: "In order that a manner of life so well adapted to the peculiarities of capitalism could be selected at all, i.e., should come to dominate others, it had to originate somewhere, and not in isolated individuals alone, but as a way of life common to whole groups of men. This origin is what really needs explaining."[9]

Of course, part of such an explanation would have to be concerned with the literal whipping of the labor force into shape, but Weber leaves this aspect of the explanation to others. The approach Weber takes is to focus on developments in religious ideas. At the outset, it should be emphasized that Weber's claim is not that certain religious ideas clandestinely maintain their hold over modern entrepreneurs, managers, and workers, keeping them within the bounds of the established order. On the contrary, Weber acknowledges that there is no longer any need for "transcendental sanctions"[10] because the modern economic order is so closely bound to the conditions of the mechanized production process that it determines the lives of individuals in this order with "irresistible force."[11] (This compulsion of individuals by the economic order is only one facet of the iron cage. Later I will discuss another important feature of the cage.)

So Weber's claim concerning religious ideas is not that they still play a role in modernity, but that religious ideas did play a role—early on—in the creation of individuals who eagerly take their positions in the modern economic processes of production and consumption. According to Weber, the Protestant Reformation exerted this influence on the development of modern individuals. What Protestantism accomplished was the reversal of the Catholic attitude toward worldly activity, which Weber describes as indifference.[12]

In my criticism of Arendt's interpretation of Christianity, I touched on the attitude toward earthly life expressed in the gospels of the New Testament and in the writings of Augustine. I described this attitude as an ambivalence toward mortal, earthly life, not as an indifference. The point of that description was to indicate another pole in the range of Christian attitudes toward earthly life besides the reverence for such life which Arendt identified as the defining feature of Christianity. For Weber's purposes, however, the Catholic attitude toward earthly life may best be characterized as indifference, not ambivalence. I have no quarrel with this characterization, since Weber contrasts this indifference with any positive evaluation, or reverence, of life, and therefore seems to support my argument against Arendt.

The indifference Weber emphasizes is found in the gospels (e.g., Matthew 6:25; Luke 12:22–3), but the clearest source for it is one of the epistles of St. Paul. For Paul, earthly, mortal life was not something to be hated (see John 12:25) or lost (see Mark 8:35). In his first epistle to the Corinthians, Paul advises the faithful not to hate this life but to continue in their earthly activity while waiting for the second coming of Christ. Noting that humanity's time on earth had been shortened (I Corinthians 7:29) and that the form of this world was passing away (I Corinthians 7:31), Paul advised that "each man remain with God in that *condition* in which he was called" (I Corinthians 7:24; also see verse 20). The second coming was almost at hand,

Paul taught, so people should not be unduly concerned with their situation in this world.

It is this "Pauline indifference" which figures prominently in Weber's treatment of Catholicism in *The Protestant Ethic.* According to Weber, Protestantism eventually reversed this Catholic indifference to life on earth and came to positively value earthly activity. And Protestantism accomplished this feat, in part, by reinterpreting the idea of a 'calling' which is found in the preceding quote from Paul. Although Weber acknowledges that there were "certain suggestions" of such a positive evaluation in the Middle Ages and in classical Greece (e.g., see my discussion of Xenophon's *Oeconomicus* in Chapter 2, pp. 31–33), he claims that the Reformation brought something "unquestionably new" to the positive estimation of worldly activity: "the valuation of the fulfillment of duty in worldly affairs as the *highest* form which the moral activity of the individual could assume" (emphasis added)[13]. And this revaluation of the calling had important consequences for the development of capitalism, according to Weber.

This Protestant revaluation, of course, did not occur all in a moment, and the Puritanism upon which Weber's argument chiefly rests is distinct in important ways from the writings of the sixteenth-century reformers. In setting up Weber's argument concerning Protestantism and capitalism, I would like to spend a little more time examining the changes within this revaluation than Weber himself does.

LUTHER AND CALVIN'S ATTITUDES TOWARD EARTHLY LIFE

This shift in attitude toward worldly activity began, as Weber indicates, with Martin Luther's own transformation in regard to worldly activity. At first, Luther's position was very much the same as that expressed in Paul's first letter to the Corinthians. In Luther's commentary on the seventh chapter of that epistle, which he wrote in 1523, Luther emphasizes Paul's message "that all outward things are optional or free before God and that a Christian may make use of them as he will; he may accept them or let them go."[14] A little later in that commentary, Luther gets more specific about the proper Christian relation to things of this world, and offers some advice on how a Christian can maintain an indifferent attitude toward mortal life. In reference to verses 29–31 of Paul's letter, which I have mentioned, Luther writes:

> This is a general teaching of all Christians, that they should treasure that eternal blessing which is theirs in the faith, despising this life so that they do not sink too deeply into it either with love and desire or suffering and boredom, but should rather behave like guests on earth, using everything for a short time because of need and not for pleasure.... A Christian should hold to this principle also in all other things. He should only serve necessity and not be a slave to his lust and nurture his old Adam.[15]

Along with this indifference to worldly activity, Luther also shared with Paul (at least early in his career) that interpretation of the calling which was expressed in the letter to the Corinthians. Paul advised the Corinthians to remain as they were when they were called, because that condition in which one was called made no difference to one's salvation. Whether one was married or not, circumcised or not, a slave or not, made no difference.[16] In his commentary on Corinthians, Luther reiterates this indifferent attitude toward the calling as the condition in which one was called and expands the scope of Paul's examples:

> And what Paul here says concerning a slave [that he should not mind, or care about, his bondage], the same is to be said of all paid servants, maids, day laborers, workmen, and domestics in their relations to their masters and mistresses. It should also be said of all vows, associations, corporations, or any tie by which one person is related or obligated to another: in all these matters service, loyalty, and duty are to be maintained, regardless of whether the one party is Christian or non-Christian, good or bad, so long as they do not hinder faith and justice and allow you to live your Christian life. For all such estates are free and no impediment to the Christian faith.[17]

Luther's interpretation of the calling, however, gradually moved away from the indifference of St. Paul, and worldly activity came to be a matter of great concern to him. The performance of one's calling in this world became a positive duty, and was no longer "optional and free before God." Weber notes this shift in Luther's thought and describes it as an increasing "traditionalism," by which Weber means the maintenance of traditional economic relations.[18]

But in any case, this increasing traditionalism of Luther's thought is not of great concern to Weber. In fact, he claims that the chapter of *The Protestant Ethic* which concerns Luther's conception of the calling was only meant to determine that the Lutheran sense of the calling "is at best of questionable importance" for the questions which concern Weber.[19] From Weber's perspective, what is significant about the idea of the calling is the way in which it eventually came to challenge traditionalism, the way in which it helped to usher in the modern economic order. And in this regard it is the interpretation of the calling which was developed by various Calvinist sects, not Luther, which is important. As a result, Weber does not offer much of a description of the traditionalism of the later Luther or an explanation for this shift.[20]

For my purposes, it is worthwhile to examine Luther's later notion of the calling in greater detail. The way Weber leaves it, one does not get any significant impression of the disciplinary dimension of Luther's conception of the calling.[21] A brief examination of some of Luther's, as well as Calvin's, writings on worldly activity and the calling, however, will help to uncover the disciplinary impulse behind that Puritan conception of the calling which Weber emphasizes and also help to bring out the aspect of Weber's argument which most interests me—the creation of modern producer/consumers, not entrepreneurs.

In 1525, the peasants who were revolting in the German territory of Swabia issued a list of their demands in the form of a pamphlet called "The Twelve Articles." In the same year, Luther responded to these demands in his "Admonition to Peace."[22] After identifying the princes and lords of Germany as the source of the rebellious temper of the German peasantry and chastising them for this, Luther addressed the peasants themselves. He criticized those peasants for revolting against the oppressive situation in which they found themselves and was especially perturbed at the peasants' use of the gospel to justify their rebellion. Luther found that everything in the articles of the peasants "is concerned with worldly and temporal matters." He replied:

> You want power and wealth so that you will not suffer injustice. The gospel, however, does not become involved in the affairs of this world, but speaks of our life in the world in terms of suffering, injustice, the cross, patience, and contempt for this life and temporal wealth.... Therefore, you must take a different attitude. If you want to be Christians and use the name Christian, then stop what you are doing and decide to suffer these injustices.[23]

With the outbreak of the Peasants' War of 1525, however, Luther abandoned the patient chastisement of the peasants he had expressed in the Admonition and called for the swift and violent suppression of the rebellion. In a pamphlet entitled "Against the Robbing and Murdering Hoardes of Peasants," Luther advised leaders, both Christian and heathen, to take up the sword against the peasants, who were "robbing and raging like mad dogs."[24] It was the duty of princes and rulers—and all other Christians as well—to destroy the peasants precisely for their rebelliousness. Luther says of the revolting peasant:

> anyone who can be proved to be a seditious person is an outlaw before God and the emperor; and whoever is the first to put him to death does right and well. For if a man is in open rebellion, everyone is both his judge and executioner; just as when a fire starts, the first man who can put it out is the best man to do the job. For rebellion is not just simple murder; it is like a great fire, which attacks and devastates a whole land. Thus rebellion brings with it a land filled with murder and bloodshed; it makes widows and orphans, and turns everything upside down, like the worst disaster. Therefore let everyone who can, smite, slay, and stab, secretly and openly, remembering that nothing can be more poisonous, hurtful, or devilish than a rebel.[25]

So, for Luther in 1525, the calling was hardly a thing of indifference. It was the duty of the peasants to endure their oppressive situation. When the peasants rejected their calling and rebelled against their worldly condition, they "abundantly merited death in body and soul,"[26] and it then became the duty of all Christians to execute this sentence. During this rebellious period, Luther's conception of the calling was directed primarily at maintaining order in the affairs of this world. It was the duty of everyone to maintain the existing order; one's salvation depended on it.

Later in his career, Luther continued to use the idea of the calling as an instrument of order. In his lectures on Genesis, which were written in the mid-1530s,[27] Luther still maintained that there was a duty to perform one's calling, but he offered a different argument in support of this claim than the one he offered to the peasants. No longer was it a matter of the duty to bear one's burdens on earth; rather, it was a question of maintaining the order that God had created on, or as, earth.

In one of these lectures, Luther claims that to abandon one's calling, or even to change one's calling, without direction from God or one's "superiors," is a sin. In his interpretation of Genesis 16:9, where an angel tells Hagar, the runaway maidservant of Sarah, to return to her mistress, Luther concludes:

> Therefore no one should change his position in life because of his own judgment or desire. God will change it either through death or because of the desire and judgment of those who are your superiors. If this does not happen, those who give up their vocations commit a sin.[28]

If one is to avoid sin, therefore, one must remain in and fulfill the position in which God has called one, no matter how unpleasant or servile that calling might be. To make this teaching less difficult to bear, Luther claims that in the eyes of God, all callings are alike. He emphasizes that this last point about the equality of all callings before God "must often be impressed upon men, for it makes hearts confident and prevents the dangerous abandoning of a calling, the abandoning that is never attempted without sin."[29]

The conception of the calling which Luther held later in his career was not so much traditionalistic, as Weber claims, in the sense that it was directed toward maintaining traditional economic relations, as it was disciplinary or regulatory. Luther was not so much concerned with particular relations as he was with order itself. Luther left open the possibility that God might use "superiors" to change things, but such changes would be orderly, that is, according to God's ordered power. This disciplinary conception of the calling was also present in the writings of Jean Calvin, the other great reformer of the sixteenth century. And it was Calvin, more than Luther, who influenced those Puritan sects so important for Weber. Nevertheless, Weber spends even less time on Calvin's conception of the calling than he does on Luther's. This is because Calvin's influence on the Puritans was centered not on his notion of the calling, but on another concept—predestination—which I will get to shortly. But the disciplinary dimension of Calvin's idea of the calling must be briefly indicated here, because it is quite different from the idea of the calling that Weber finds among the Puritans.

In the third book of his *Institutes of the Christian Religion*, which he wrote in 1535 or 1536, Calvin sounds very much like Luther did in that same period in regard to the calling. Calvin says of the calling: "He only who directs his life to this end will have it properly framed; because, free from the impulse of rashness, he will not attempt more than his calling justifies, knowing that it is unlawful to overleap the prescribed bounds."[30] For both these reformers, the calling served as a call to order and a guarantee that order would be maintained. For Luther, the idea of the calling required that "no one change his position in life because of his own judgment or desire," and for Calvin, it required that no one "overleap the prescribed bounds."

And while Luther emphasized, as a sort of ministerial pointer, the value of the idea that all callings were equal before God, Calvin makes a similar point, although he does not claim all callings are equal. Calvin writes:

> Again, in all our cares, toils, annoyances, and other burdens, it will be no small alleviation to know that all these are under the superintendence of God.... Every one in his particular mode of life will, without repining, suffer its inconveniences, cares, uneasiness, and anxiety, persuaded that God has laid on the burden. This, too, will afford admirable consolation, that in following your proper calling, no work will be so mean and sordid as not to have a splendour and value in the eye of God.[31]

Even if all callings are not equal in the eye of God, they are all of value, according to Calvin; this should console those who find themselves in unpleasant or oppressive circumstances, and help them to avoid the sin of rebellion or unauthorized assertion.

The Protestant reversal of the Pauline indifference toward earthly life was not limited to the disciplinary development of Luther's and Calvin's conceptions of the calling, however. There is another facet of this reversal I would like to emphasize before moving on to Weber's discussion of Puritanism. Like the disciplinary character of their conceptions of the calling, the feature of Luther's and Calvin's thought that I now want to stress is one which Weber passes over in *The Protestant Ethic*, but it too is important for the claims I will eventually make concerning Protestantism and the value of convenience.

In regard to worldly activity in general, beyond the requirements of one's calling, Calvin, and eventually Luther, took positions which cannot be described as indifferent. Early in Luther's career, when he wrote his commentary on I Corinthians 7, he advised Christians to "behave like guests on earth, using everything for a short time, because of need and not for pleasure." By following this lesson from Paul's epistle, one could avoid sinking "too deeply" into earthly life. But as Luther's conception of the calling moved away from Paul's indifferent attitude toward the particular situation in which Christians were called, so too did his estimation of worldly activity in general. By the mid-1530s, when he was writing his lectures on Genesis, Luther saw more to earthly activity than the mere satisfaction of necessity; it was something joyful, to be enjoyed. And for his part, Calvin also rejected advice such as that of the early Luther and found earthly activity to be enjoyable—within moderation of course.

The point I want to make, however, is not just that Luther and Calvin came to see earthly life as something to be enjoyed. Alongside this evaluation, there also occurred in the thought of these reformers an intensification of that contempt for earthly life which I identified earlier in challenging Arendt's interpretation of Christianity. What I want to stress is neither the less severe side of Luther's and Calvin's evaluation of earthly life nor their contempt for such life. I want to emphasize the intensified ambivalence of their thought. The ambivalence of these reformers was not an indifferent, either this way *or* that, attitude, as one might expect from the current usage of this word; their ambivalence was an emphatic this way *and* that.[32]

In Luther's thought, this ambivalence is found quite distinctly in his lectures on that section of Genesis which I stressed in my argument with Arendt (i.e., Genesis 3:16–9), in which God punishes Adam and Eve for eating from the tree of knowledge. On the one hand, Luther expounds upon the punishments and fills in the broad parameters which were established in the verses themselves. For instance, in regard to woman's punishment (verse 16), Luther claims that "Eve's sorrows, which she would not have had if she had not fallen into sin, are to be great, numerous, and of various kinds."[33] Woman's punishment is not just the increased pain in childbirth, which is mentioned in the verse, but also the "severe and sundry ailments" women may suffer throughout pregnancy, which are listed by Luther,[34] as well as the "various dangers" a woman encounters "during all the rest of her life, while she devotes herself to her children."[35] And besides the procreative dimension of the punishment, Eve, who prior to her sin "was very free and...was in no respect inferior to her husband," became subject to his rule.[36]

Now on the other hand, despite this grim portrayal of the earthly life of woman, Luther also claims that Eve's punishment was "truly happy and joyful."[37] And this is not just because God made it possible for her to attain eternal life through this punishment, although this undoubtedly is the most joyful aspect of earthly life for Luther. But even on earth, the life of woman is joyful. In her punishment, Eve sees that

> she is not being deprived of the blessing of procreation, which was promised and granted before sin. She sees that she is keeping her sex and that she remains a woman. She sees that she is not being separated from Adam to remain alone and apart from her husband. She sees that she may keep the glory of motherhood.[38]

Furthermore, motherhood is joyful not just for women, but for men as well. Part of the glory of motherhood, claims Luther, "is that we are all nourished, kept warm, and carried in the womb of our mothers; that we nurse at their breasts and are protected by their effort and care."[39] Expressing his personal enjoyment of motherhood, Luther writes: "To me it is often a source of great pleasure and wonderment to see that the entire female body was created for the purpose of nurturing children. How prettily even little girls carry babies on their bosom."[40]

For Luther, therefore, motherhood was both a punishment and a blessing, a sign of God's wrath and of his mercy, a source of guilt and hope. Luther displayed a similar ambivalence in regard to Adam's punishment (verses 17–9). But here, Luther not only expands the list of punishments, as he did with Eve's, but claims that they have gotten more severe since the time of Adam. Luther points out that, according to Genesis, Adam had only to contend with the misfortunes of "thorns, thistles, and hard work"[41] in providing food for the household.

> But now we learn from experience that countless others have been added. How many kinds of damage and how many diseases affect the crops, the plants and the trees, and finally everything that the earth produces! Furthermore, frosts, lightning bolts, injurious dews, storms, overflowing rivers, settling of the ground, earthquakes—all do damage.[42]

That the list of calamities which could befall a farmer had increased since Adam's time, Luther interpreted as a consequence of the increasing corruption and seduction of humans by Satan. Luther claimed to be "fully of the opinion that because of the increase of sins the punishments were also increased and that these troubles were added to the curse of the earth."[43] But it was not just the increase in the variety of disasters which convinced Luther that the world was becoming more depraved. Luther also believed that farmers in his day experienced "more frequent disasters to crops than in former times," and that this was another indication that "the world [was] deteriorating from day to day."[44]

One might imagine that the subjection of Eve, while being a punishment for women, was a blessing for men, but Luther interprets this subjection as a punishment for both sexes. The husband's rule over his wife and the household "cannot be carried on without the utmost difficulties."[45] And as was the case with the agricultural burdens which were placed upon man, Luther claims that the tasks of "supporting, defending, and ruling over his own...are far more difficult in our age than they were in the beginning."[46] This was due, of course, to the increasing "perversity of people."

On the one hand, therefore, Luther saw the life of the common householder as being a sign of condemnation by God. All human activity bears witness to the fall. As Luther puts it:

> whenever we see thorns and thistles, weeds and other plants of that kind in a field and in the garden, we are reminded of sin and the wrath of God as though by special signs. Not only in the churches, therefore, do we hear ourselves charged with sin. All the fields, yes, almost the entire creation is full of such sermons, reminding us of our sin and of God's wrath, which has been aroused by our sin.[47]

But on the other hand, just a few pages later in his lecture, Luther claims that, although farmers "are plagued with hard labor, that labor is seasoned with matchless pleasure, as daily the new and wonderful sight of the creatures impresses itself upon their eyes."[48] The life of the husband, therefore, like the life of the wife, is at once both a reminder of the fall from grace and a source of pleasure.

Luther's interpretation of both Adam's and Eve's punishment displays an ambivalence toward earthly life. Luther expounds upon the punishments God inflicted upon men and women in their earthly lives, but he also takes pains to point out the pleasurable, enjoyable aspects of earthly life. At the beginning of his lecture on Genesis 3:16–9, Luther emphasizes that the "godly" must not despair at the severity of mortal life but must turn "to the outside what is beautiful." This "means not merely looking at what is evil but delighting in God's gifts and blessings and also burying the punishments, annoyances, pains, griefs, and other things."[49]

Calvin's ambivalence toward earthly life took a different form from Luther's. Like Luther, Calvin recognized that human life, since Adam's and Eve's fall from grace, was plagued by various scourges and calamities, but he did not interpret these vexations as a sign that people were becoming increasingly depraved. There are none of Luther's eschatological premonitions in Calvin's interpretation of Adam's and Eve's punishments. For Calvin, these punishments were not a sign that another disaster on the order of the biblical Deluge was in store, as they were for Luther.[50] Instead, Calvin saw in the many vexations and annoyances of life God's reminder to men and women that the things of this life on earth were uncertain and fleeting. Describing God's response to human folly, Calvin writes:

> In short, the whole soul, ensnared by allurements of the flesh, seeks its happiness on the earth. To meet this disease, the Lord makes his people sensible of the vanity of the present life, by a constant proof of its miseries. Thus, that they may not promise themselves deep and lasting peace in it, he often allows them to be assailed by war, tumult, or rapine, or to be disturbed by other injuries...by diseases and dangers he sets palpably before them how unstable and evanescent are all the advantages competent to mortals.[51]

In order for Christians to keep their attention on the eternal, truly happy life which awaits them in heaven, they must refrain from becoming too deeply involved with the affairs of earth. Calvin's advice to Christians in this regard is not to become indifferent to earthly life, to be able to either accept the things of this world or let them go, as the early Luther, following Paul, had advised; instead, he takes a much stronger position. Calvin claims "that our mind never rises seriously to desire and aspire after the future, until it has learned to despise the present life."[52] For Calvin, "there is no medium between the two things: the earth must either be worthless in our estimation, or keep us enslaved by an intemperate love of it."[53]

Given these rather extreme statements, one might expect that Calvin would have adopted the early Luther's distinction between pleasure and necessity, or the Augustinian distinction between use and enjoyment (see note 15 of this chapter) and recommended that Christians limit their involvement in earthly life to necessity only. But Calvin rejects such a recommendation as being "unnecessarily austere"[54] and, in his rejection, appears to contradict his earlier claim that "there is no medium between" the worthlessness of life and the enslavement to it. In the chapter immediately following the one in which the claim against a medium was made, Calvin writes:

> For if we are to live, we must use the necessary supports of life; nor can we even shun those things which seem more subservient to delight than to necessity. We must therefore observe a mean, that we may use them with a pure conscience, whether for necessity or for pleasure.[55]

So on the one hand, Calvin says that in order to avoid becoming enthralled by earthly pleasures, there can be no mean; life on earth must be worthless to Christians. On the other hand, he says that earthly delights can not be avoided, so Christians must strike a mean between necessity and pleasure. If Calvin had left things as they stand here, one could explain the apparent contradiction of these claims about earthly life by pointing out that the former claim establishes an ideal, while the latter claim is made in a spirit of concession to reality. Ideally, Christians should be contemptuous of life on earth, but since such a life does have its fleeting pleasures and delights, Christians should strive to maintain a balance in their enjoyment of things of this life.

But Calvin does not leave things as they stand here. He does more than simply concede that earthly life is pleasurable; he points out how all creation is intended to be pleasurable to men and women. "There is not one little blade of grass, there is no color in this world that is not intended to make men rejoice,"[56] claims Calvin. Consequently, Christians are "are not only to be spectators in this beautiful theater but to enjoy the vast bounty and variety of good things which are displayed to us in it."[57]

As a guiding principle for such enjoyment of earthly things, Calvin suggests that Christians "refer them to the end for which their author made and destined them, since he created them for our good, and not for our destruction."[58] He then gives some examples to elucidate this point:

> Now then, if we consider for what end he created food, we shall find that he consulted not only for our necessity, but also for our enjoyment and delight. Thus in clothing, the end was, in addition to necessity, comeliness and honour; and in herbs, fruits, and trees, besides their various uses, gracefulness of appearance and sweetness of smell.[59]

Given this attitude toward pleasure, Calvin can hardly be considered an extreme ascetic, and in fact, he described as "inhuman" that philosophy which would restrict human earthly activity to the satisfaction of necessity alone.[60] Therefore, the severe "worldly asceticism" of the Puritans which Weber emphasizes in *The Protestant Ethic*, and which I will discuss in the next section of this chapter, cannot be derived from Calvin's teachings on earthly pleasures.[61]

But Calvin, of course, was no hedonist, either. Christians should enjoy food and drink, fine clothes and flowers, and all other things which please the senses,[62] but such enjoyment should never be carried to the point where it distracts them from a pious concern for the true, eternal happiness which is promised to Christians, or from the fulfillment of one's calling.[63] Here, too, Calvin offers some advice for "curbing licentious abuse" of earthly pleasures: "There is no surer or quicker way of accomplishing this than by despising the present life and aspiring to celestial immortality."[64]

While this advice may appear to be another contradiction within Calvin's thought—in that his advice amounts to telling Christians to enjoy life's pleasures, but to curb excesses by despising this life—I prefer not to describe this antinomy as such. At least since Hegel, the idea of a contradiction seems to imply the immanent resolution of the tension involved therein, but for Calvin, as for Luther, the tension between the enjoyment of, and the contempt for, life on earth was to be maintained. Both these reformers rejected the asceticism of monastic orders, which can be interpreted as one route for eliminating such tension. Instead, they heightened the tension between these two attitudes toward earthly life—Luther by drawing out the dark, punitive dimension of that life while nonetheless insisting on "turning to the outside what is beautiful" and Calvin by indicating how pleasing all of creation was meant to be while insisting that Christians despise earthly life.[65] As I have already indicated, I prefer to describe this tension by the term 'ambivalence.'

The heightened ambivalence of Luther and Calvin could not long be maintained by the followers of Protestantism, however, and Weber's notorious text examines the different evaluation of earthly life which Protestantism eventually established. Before more closely examining Weber's argument, I should point out that the loss of Christian tension or ambivalence in regard to earthly life helped to cultivate the modern value of convenience. But the shift from ambivalence (or as Weber describes it, indifference) to a comfortable, convenient life in the cage was not a simple, straightforward process.

WORLDLY ASCETICISM AND THE EMERGENCE OF THE CAGE

Having pointed out both the ambivalence of Luther's and Calvin's attitudes toward earthly life and the disciplinary nature of their conceptions of the calling, I can now pick up Weber's argument relating Protestantism, specifically Puritanism, and capitalism. Eventually I will return to these points about Luther and Calvin, and situate them in regard to Weber's argument. But first, I must discuss another of Calvin's doctrines—predestination—the one which Weber claims was the "most characteristic dogma" of Calvinism.[66] The doctrine of predestination is crucial to the relation between Protestantism and capitalism which Weber tries to establish in *The Protestant Ethic.* In the *Institutes*, Calvin describes this doctrine as follows:

> By predestination we mean the eternal decree of God, by which he determined with himself whatever he wished to happen with regard to every man. All are not created on equal terms, but some are preordained to eternal life, others to eternal damnation; and, accordingly, as each has been created for one or other of these ends, we say that he has been predestinated to life or death.[67]

This doctrine did not originate with Calvin, however. In his letter to the Ephesians, Paul had expressed the idea that God, "before the foundation of the world," chose those who would be saved (Ephesians 1:4–5),[68] and to the Romans, Paul wrote, "For whom He foreknew, He also predestined *to become* conformed to the image of His Son" (Romans 8:29). Although the letter to the Ephesians, as well as the preceding quote from Romans, refer only to the predestination of those who were saved, Paul does give some indication in his letter to the Romans that others were predestined to damnation. At least he leaves open this possibility: "What if God, although willing to demonstrate His wrath and to make His power known, endured with much patience vessels of wrath prepared for destruction? And he did so in order that he might make known the riches of His glory upon vessels of mercy, which He prepared before hand for glory..." (Romans 9:22–3).[69] Paul almost seems to suggest that the predestination of those who are saved required the predestination of those who are damned, so that there would be a contrast. In any case, Calvin interprets Paul's statements on predestination as including both those who are saved and those who are damned.

Another important feature of this doctrine, one which Weber emphasizes, is that it effectively places the salvation of an individual beyond the influence of that individual or the church. God determined whether one would be saved or damned before he ever created humans, and there was nothing one could do to alter this situation. Calvin insisted that one was predestined by God "according to the good pleasure of his will...," and "wherever this good pleasure of God reigns, no good works are taken into account."[70] Paul expressed this gratuitous dimension of the doctrine of predestination in his letter to the Romans,[71] and Augustine, writing in the fifth century, also insisted that the salvation of the elect was gratuitous.[72] But in the course of those years in which the Roman Catholic Church established itself as the dominant religion of the Western world, the idea that there was nothing that one could do to attain salvation became buried under the proliferation of sacraments, some of which served as steps one could take to ensure one's salvation (e.g., penance and extreme unction).

Protestantism, of course, rejected most of these sacramental developments, and Calvin's conception of gratuitous predestination was, in some sense, a weapon to be used in this struggle with the Catholic Church. For example, Calvin was sharply critical of the sacrament of penitence, or penance, as it was established by the "schoolmen." According to the Catholic sacrament of penitence, the repentant had to perform satisfaction for their sins in some manner: prayers, fastings, gifts to the church or to the poor, or other charitable works could be required of the sinner. Besides these forms of satisfaction there arose the practice of purchasing indulgences. According to this practice, one could buy from the church a pardon for the satisfaction that one had to perform.

Calvin was thoroughly revolted by this crass practice, in which "the salvation of the soul [was] made the subject of a lucrative traffic, salvation taxed at a few pieces of money, nothing given gratuitously."[73] However, the selling of indulgences was for Calvin only the most obvious form of the blasphemy which the church had committed by requiring satisfaction to be performed for sins. The very idea of satisfaction being required for forgiveness amounted to the purchase of salvation. Calvin summarizes the papal attitude toward this element of penitence as follows:

> They say that it is not sufficient for the penitent to abstain from past sins, and change his conduct for the better, unless he satisfy God for what he has done; and that there are many helps by which we may redeem sins, such as tears, fastings, oblations, and offices of charity; that by them the Lord is to be propitiated; by them the debts due to divine justice are to be paid; by them our faults are to be compensated; by them pardon is to be deserved.[74]

According to the doctrine of predestination, however, the forgiveness of sins is free, *gratis*. It is not just that one does not have to pay for one's sins to be forgiven; one cannot pay. "Assuredly," insists Calvin, "divine grace would not deserve all the praise of election, were not election gratuitous; and it would not be gratuitous, did God in electing any individual pay regard to his future works."[75]

It was not only the sacrament of penance that suffered at the hands of Calvin, however. The doctrine of predestination, when carried to its logical conclusion, seemed to undermine the importance which the Catholic Church attributed to the sacraments in general. The position the mature Church took in regard to the sacraments is exemplified in the writing of Thomas Aquinas, who wrote in the thirteenth century. In his *Summa Theologica*, Aquinas recognized seven sacraments, and argued that these sacraments were necessary for salvation, and were "instituted by God to be employed for the purpose of conferring grace."[76]

Although Calvin does not address Aquinas explicitly in regard to the issue of the sacraments, he does challenge these Catholic teachings. To begin with, Calvin recognizes only two of the Church's seven sacraments—baptism and eucharist—and spends one of the last chapters of the *Institutes* denying the legitimacy "of the Five Sacraments, falsely so called."[77] But even the two sacraments Calvin recognized were not necessary for salvation. "Assurance of salvation," writes Calvin, "does not depend on participation in the sacraments, as if justification consisted in it. This, which is treasured up in Christ alone, we know to be communicated, not less by the preaching of the Gospel than by the seal of the sacraments, and may be completely enjoyed without this seal."[78] And as for the teaching that grace is conferred by the sacraments, Calvin replies; "They confer nothing."[79]

Even though Luther, along with Calvin, professed the doctrine of predestination, Weber indicates that the doctrine waned in importance for Luther "the more his position as responsible head of his Church forced him into practical politics."[80] For Calvin, however, the doctrine assumed a central role as his teaching developed. This difference helps to explain why Calvin broke more cleanly with the Catholic Church and its sacraments than did Luther.[81] According to Weber, Luther maintained that God's saving grace could be lost through sin and recovered through penitent humility and participation in the sacraments.[82] On the other hand, Calvin, who clung firmly to the doctrine of predestination, rejected the possibility of either losing or recovering grace. Consequently, claims Weber, the practice of private confession disappeared "from all the regions of fully developed Calvinism."[83]

For Weber, the disappearance of this sacrament was "an occurrence of the greatest importance.... The means to a periodical discharge of the emotional sense of sin was done away with."[84] Weber's point, as I read him, is that this emotional energy which had been discharged in, or generated and regulated by, the sacrament of confession, could now be focused in a different direction by Calvinists. But there is more to the doctrine of predestination than the gratuitousness of salvation and the corresponding challenge to the Catholic sacraments. There is another important element of Calvin's doctrine of predestination which must be mentioned here, what might be called the gratuitousness of damnation. This other dimension of predestination posed its own particular threat to Catholic doctrine, and generated its own emotional energy distinct from the guilt for sin. It was this element of the doctrine of predestination, not the gratuitousness of salvation, which seems to have most troubled the Catholic Church.

For despite the implicit antagonism between the idea of gratuitous salvation and the Catholic attitude toward the sacraments, the Church nevertheless accepted the doctrine of predestination, particularly the idea of gratuitous salvation. Aquinas, who a few paragraphs ago was cited to indicate the importance Catholic theology placed on the sacraments, can also be used to present the Church's view of the doctrine of predestination. According to the *Summa Theologica*:

> God wills to manifest His goodness in men; in respect to those whom He predestines, by means of His mercy, in sparing them; and in respect of others, whom He reprobates, by means of His justice, in punishing them.... why He chooses some for glory and reprobates others has no reason except the divine will.[85]

In this passage, Aquinas, the "angelic Doctor" of Catholicism, sounds somewhat like Calvin, the heretic. Both agree that salvation is predestined for some and that it depends on the goodness and mercy of God. Of course, for Aquinas to profess this element of predestination and also claim that the sacraments were necessary for salvation, he had to resort to some rhetorical maneuvers to get around the conflict between these ideas. For Calvin, such maneuvers were nothing more than sophistical subterfuges, and he attacked them as such.[86] What I want to stress, however, is not the way the Catholic Church accommodated these problematic ideas or the way Calvin sought to upset that accommodation; instead I want to point out that element of the doctrine of predestination that the Catholic Church could not accept.

The preceding quote from Aquinas gives some indication of the exclusion I want to emphasize. In that quote, Aquinas uses the term predestination in reference only to the saved; the term reprobation is used in regard to the damned. For Aquinas, predestination does not include those who are not saved; reprobation, as something distinct from predestination, is the source of their damnation. This distinction is made more clearly in the following quote, and its implications are more fully drawn out by Aquinas (and myself):

> Thus, as predestination is a part of providence, in regard to those *divinely ordained to eternal salvation*, so reprobation is a part of providence in regard to those *who turn aside* from that end.... Therefore, as predestination includes the will to *confer*

> grace and glory, so also reprobation includes the will to *permit* a person *to fall into sin*, to *impose the punishment* of damnation *on account of that sin.*[87] (Emphasis added.)

As I have tried to indicate through the use of italics, predestination, without merit, belongs only to the saved; upon them grace and glory are conferred; they are ordained to eternal salvation. The damned, on the other hand, turn aside from eternal salvation; God permits them to fall into sin, but does not ordain that they do so. And they are punished on account, or because, of their sin.

Aquinas, in other words, lets God off the hook for the damnation of the non-elect, and shifts responsibility to the damned themselves. Ultimately, of course, it is God's will which permits them to sin, but the damned earn their damnation through their sins. The responsibility established by this distinction between predestination and reprobation can be interpreted as the general, broad form of that specific responsibility which was institutionalized by the Church in the thirteenth century, when the sacrament of confession became mandatory.[88] (In order to be forgiven by the priest, one had to be responsible for one's sins.) But the Catholic rejection of the gratuitousness of damnation did not begin with Aquinas in the thirteenth century. Rather, the doctrine of predestination caused problems early on in the formation of Church dogma.

As I indicated earlier, Paul, in his letter to the Romans, can be read as including both the saved and the damned under the notion of predestination. Following Paul, Augustine, in the fourth and fifth centuries, believed that the majority of people were predestined to damnation[89] and borrowed Paul's imagery of the "vessels of wrath" to explain this point.

> The other mortals, not of this number [of the elect], who are of the same mass as these, but have been made vessels of wrath, are born for their advantage. God creates none of them rashly or fortuitously, and He also knows what good may be made from them.[90]

Even though Augustine defended the justice of God in his doctrine of predestination by placing responsibility for human punishment on Adam[91] and explicitly rejected one of the doctrine's logical, yet troubling, conclusions (i.e., that sinners should not be admonished or punished, since they may have been predestined to be among the damned[92]), Augustine's doctrine nevertheless became a source of conflict within the Church during the century following his death.[93] The semiPelagians, who tried to strike a compromise between the conflicting teachings of Pelagius and Augustine (see note 72), rejected the doctrine of double predestination, that is, that both the saved and the damned were predestined.[94] During the second half of the fifth century, the semiPelagians were successful in having the doctrine of double predestination condemned by regional synods, particularly in Gaul, or France.[95] But it was at the second Council of Orange, in 529, that the followers of Pelagius were most successful in this particular battle with Augustine.

Overall, Augustinianism fared much better than semiPelagianism did at this official proclamation of church teachings; the twenty-five propositions or canons to which the members of the Council subscribed were drawn largely from the writings of Augustine himself.[96] But in the five-point creed which the signatories added to the document, the Augustinian doctrine of predestination, or at least the disturbing half of it, was excluded from the Church's official theology. Actually, only two of the five points were involved in this mutation. After indicating how Adam's sin weakened the will of men and women so that none thereafter could love God in a suitable manner, the creed continues:

> b) All, however, are able, after they have received grace through baptism, with the co-operation of God, to accomplish what is necessary for the salvation of their soul.
> c) It is in no way our belief that some are predestinated by God to evil (predestination heresy); rather, if there are any who believe a thing so evil, we, with horror, say anathema.[97]

The teachings of the Council of Orange, with their predominantly Augustinian bent, were quite influential in the development of medieval theology,[98] but they also mark the point where the Church abandoned Augustine's idea of predestined damnation.[99] It was against this longstanding position of the Catholic Church that Calvin threw himself with all his weight. To those who urged that the doctrine of predestination as developed in

Paul's letters and Augustine's literary artillery was too troubling and dangerous a doctrine to merit widespread discussion, Calvin replied:

> There is nothing in the allegation that the whole subject is fraught with danger to pious minds, as tending to destroy exhortation, shake faith, disturb and dispirit the heart. Augustine disguises not that on these grounds he was often charged with preaching the doctrine of predestination too freely.... Those, however, who are so cautious and timid that they would bury all mention of predestination in order that it may not trouble weak minds, with what colour, pray, will they cloak their arrogance, when they indirectly charge God with a want of due consideration, in not having foreseen a danger for which they imagine that they prudently provide.[100]

Calvin insisted not only that the doctrine of predestination be taught, but that it be taught in its entirety; he rejected the truncated version of the Catholic Church, which limited predestination to election, or salvation, only.[101] In contradiction to the Catholic Church's teaching that all can be saved and none are irrevocably damned, Calvin claims: "All are not created on equal terms, but some are preordained to eternal life, others to eternal damnation; and, accordingly, as each has been created for one or other of these ends, we say that he has been predestinated to life or to death."[102]

In pushing this point about the predestination of the damned, Calvin is even more severe than Augustine. Augustine, at least, interposed the responsibility of Adam to mitigate the gratuitousness of damnation. Calvin, however, does not offer this consolation. "If we look for the source of their ruin," writes Calvin, concerning the damned, "we must ultimately come to this, that being cursed by God, all they do, say, or intend, only furthers and increases their curse. Yet, the cause of eternal rejection is so hidden that there is nothing left for us to do but to be amazed at the incomprehensible mind of God."[103] Calvin also calls this ultimate question mark the "secret" or "hidden counsel of God."[104]

And on the occasion when Calvin actually offers some sort of answer to the question of human damnation, he is hardly more comforting: "Those...whom God passes by he reprobates, and that for no other cause but because he is pleased to exclude them from the inheritance which he predestines to his children."[105] The severity of this teaching should be evident. Not only are some damned through no fault of their own, but there is no explanation for why they are damned, other than that their damnation pleases God. This is the "decretum horribile" which Calvin taught.[106]

As was the case with Augustine, Calvin's teachings on predestination became a source of controversy after his death and were the subject of several synods.[107] In Calvin's case, however, the doctrine was not shorn of its sharp edges, but was retained as a double decree which included damnation. The statement of the doctrine which Weber cites in *The Protestant Ethic* comes from the Westminster Confession of 1647, the product of one of these synods. This Confession not only states explicitly that some are "foreordained to everlasting death," but at one point claims—in direct opposition to the teachings of the Catholic Church—"All those whom God hath predestined unto life, *and those only*, He is pleased in His appointed and accepted time effectually to call by His word and spirit (out of that state of sin and death, in which they are by nature)."[108] (Emphasis added.)

I mentioned earlier that the Calvinist elimination of private confession was interpreted by Weber as a source of emotional energy. But the idea that some were hopelessly, helplessly, and gratuitously damned provided the real emotional energy of Calvinism. As Weber puts it:

> In its extreme inhumanity this doctrine must above all have had one consequence for the life of a generation which surrendered to its magnificent consistency. That was a feeling of unprecedented inner loneliness of the single individual. In what was for the man of the age of the Reformation the most important thing in his life, his eternal salvation, he was forced to follow his path alone to meet a destiny which had been decreed for him from eternity. No one could help him. No priest.... No sacraments.... No Church.... Finally, even no God. For even Christ had died only for the elect.[109]

What I would like to stress at this point, however, is not just that the doctrine of predestination threw the individual back upon himself, but also that the doctrine threw into doubt the possibility of salvation and heavenly immortality. Under Catholicism, salvation was available to all, and the sacraments provided certain steps that could be taken to almost guarantee salvation. Baptism washed away the stain of original sin and started one out on one's spiritual life. Eucharist and confirmation provided spiritual strength to help one avoid sin, and penance, or penitence, absolved one of those sins that one might nonetheless commit. And if a priest was at hand to perform extreme unction at the hour of one's death, one could be cleansed of any venial sins which one forgot or failed to confess and be "prepared for final glory."[110] In other words, the Church provided comforting answers to the question of salvation or damnation and promised immortality to those who followed its teachings and participated in its sacraments.

Calvin's doctrine of predestination radically challenged this Catholic confidence in salvation, and not just because the doctrine undermined the sacraments. For Calvin, as for Augustine, most people were from the start irrevocably denied the possibility of heavenly immortality. And it was because of this dreadful dimension of the doctrine, which threw salvation into doubt, that predestination's "psychological effect was extraordinarily powerful."[111] As Weber puts it, "The question, Am I one of the elect? must sooner or later have arisen for every believer and have forced all other interests into the background."[112]

Calvin, apparently, had no doubt about his own salvation and suspected that true believers would have the same confidence. He found it strange that "many who boast of being Christians, instead of thus longing for death, are so afraid of it that they tremble at the very mention of it as a thing ominous and dreadful."[113] For Calvin, death was not something to be avoided, either intellectually or physically; instead, he taught that Christians should "ardently long for death, and constantly meditate upon it."[114] Calvin's doctrine of predestination, I want to suggest, was effective in getting people to focus upon their eventual death, even if it did not make them long for it. The doctrine raised, in a very poignant way, the specter of human mortality.

According to Weber, however, most of Calvin's followers were unable to attain, or maintain, his confidence in heavenly immortality. "For the broad mass of ordinary men..." writes Weber, "the *certitudo salutis* in the sense of the recognizability of the state of grace necessarily became of absolutely dominant importance."[115] They needed a sign of their salvation, and one was provided in the form of the calling.

Earlier, I discussed the shift in attitude toward the calling, from the indifference of Paul's first letter to the Corinthians to the disciplinary concern of the mature Luther and Calvin. For both Luther and Calvin, the idea of a calling came to be used as a tool for maintaining order in the face of rebellious masses. But in regard to the serious questions raised by the doctrine of predestination, the idea of a calling became an instrument of change.

In their responses to the anxiety the faithful had concerning salvation, Calvinist ministers came to encourage "intense worldly activity" as the most suitable means of dispersing doubts about, and inspiring confidence in, one's salvation.[116] Weber describes this Puritanical attitude toward earthly activity and its relation to the questions raised by predestination as follows:

> It was through the consciousness that his [the Calvinist's] conduct, at least in its fundamental character and constant ideal (propositum oboedientiae), rested on a power within himself working for the glory of God; that is not only willed of God but rather done by God that he attained the highest good toward which this religion strove, the certainty of salvation.[117]

By the rigorous and conscientious performance of one's earthly calling, one not only maintained the order that God willed, as Luther and Calvin had taught, but one could "increase" or "augment" the glory of God.[118] The greater the success one had in performing one's calling, the greater the glory of God one accomplished through one's earthly activity, and the surer one could be that one was of the elect. For only one of the elect could have an "effectual calling," one capable of augmenting the glory of God with real, not apparent, good works.[119]

But it was not really success itself which was important in regard to salvation. Rather, it was the fact that one had organized one's life to serve the glory of God which was truly important, that one's life "was thoroughly rationalized in this world and dominated entirely by the aim to add to the glory of God on earth."[120] Success happened to follow upon such a dedicated, disciplined form of life and so was valuable as a sign, or proof, of one's election. At this point I should note, as Weber does, that this attitude toward success in one's calling, as providing proof of one's predestination to salvation, was quite foreign to Calvin himself. Calvin never wavered in his teaching that works have nothing to do with salvation and damnation, and he denied that works give any indication of God's "secret counsel."[121] In fact, Calvin warned against attempts to understand this realm of mystery and taught that "it is not right that man should with impunity pry into things which the Lord has been pleased to conceal within himself."[122]

Nevertheless, this deviant use to which the Puritans put the idea of the calling was, in Weber's eyes, of the greatest economic significance. It was by ascribing to the calling the important role of signifier of salvation that Calvinism was able to reclaim and redirect into economic activity that emotional energy which Catholicism had regulated through the sacraments. As individual believers successfully performed their particular callings, not only was their certainty of salvation confirmed and the glory of God enhanced, but the productivity of the community was also increased. This effect of the Puritanical use of the calling is described in utilitarian terms by Weber: "The specialization of occupations leads, since it makes the development of skill possible, to a quantitative and qualitative improvement in production, and thus serves the common good."[123]

It was this promotion of the common good that provided the basis for the distinction between real and apparent good works, and even for distinctions among effectual callings. Those callings which were useful to the community were pleasing to God, and those which were more useful than others were more pleasing to, and more greatly glorified, God.[124] According to Weber, the Puritan ministers were not opposed to changes in callings, as long as the changes were made in order to augment God's glory, that is, to increase the productivity of the community.[125] Weber's claim is that this dynamic conception of the calling, with its specializing effect, helped to usher in the rationally organized capitalism of modernity. This dynamism stands in sharp contrast to what I termed the disciplinary, or as Weber put it, "traditionalistic," conception of the calling which Luther and Calvin employed to defend the old economic order.

This is not to say that there was no disciplinary dimension to the Puritanical conception of the calling, however. As Luther's and Calvin's conceptions of the calling were aimed, in part, to restrain the rebellious activity of an expropriated peasantry, the Puritans' conception of the calling was used to fit that same stratum into the new economic order. For even though the Puritans believed that not everyone was predestined to salvation, they still believed, according to Weber, that God had prepared a calling for everyone without exception and that there was a duty to fulfill one's calling.[126] God created everyone for his glory. For those such as vagabonds and beggars who pursued no calling and were therefore a blight on the glory of God, the Puritans favored the creation of workhouses, which could instill the discipline which was required by the glory of God and the capitalist economy.[127]

Although the Puritanical notion of the calling as proof of salvation may have been a great boon to the development of modern capitalism, operating as it did to establish this order on both the level of the Puritan entrepreneur and the level of the undisciplined peasantry, the idea was nonetheless an inherently dangerous one from the religious perspective of Calvinism. For one thing, the idea that proof of one's election could be provided by the successful pursuit of one's calling came perilously close to the Catholic Church's position that one could attain salvation by good works on earth.[128] Although the Puritan ministers insisted that such works were nothing more than an indication of salvation and that salvation in no way depended on worldly success, there was always the danger that among those anxious believers who accepted Calvin's doctrine of predestination, such works would come to outweigh faith in providing certainty of salvation. Instead of bolstering the Calvinist's faith in his or her salvation, worldly success might come to replace that faith as a source of certainty.

This notion of the calling as proof also posed another threat to Calvinism, in that the success that attended the disciplined, purposeful activity of the believer could undermine that very discipline. The more successful—that is, wealthy—the Puritans became in their callings, the greater was the temptation and feasibility of living a leisurely, comfortable life. Weber points out how Richard Baxter, the Puritan minister upon whose writings he principally relies, frequently warned against the accumulation of wealth. Weber writes that for Baxter, "Wealth as such is a great danger; its temptations never end, and its pursuit is not only senseless as compared with the dominating importance of the Kingdom of God, but it is morally suspect."[129]

The primary objection Puritan ministers such as Baxter raised against wealth was that it could lead to "distraction from the pursuit of a righteous life"[130]—that is, a life devoted solely to the glorification of God. According to Weber, the Puritan ministers were more suspicious of wealth than was Calvin himself.[131] Given the greater weight those ministers placed upon worldly activity, in comparison to Calvin, it is understandable that they would be more concerned than he was with the dangers of earthly success. And in response to this threat, which followed upon their novel idea that salvation could be proven, Calvinist ministers were compelled to move even farther away from Calvin's teachings. They developed what Weber calls "worldly asceticism," something quite distinct from the ambivalence to earthly life which Calvin (and Luther) maintained.

Weber identifies two principal features of this Puritanical asceticism. The first of these is the prohibition against wasting time. For the Puritans, claims Weber, "waste of time is...the first and in principle the deadliest of sins."[132] Any time not spent performing one's calling was wasted time, since that time could have been spent in furthering the glory of God. This concern with time did not originate with the Puritans, but arose instead among those otherworldly ascetics, the Catholic monks.[133] And it was not even the Puritans who first brought this monastic regulation of time out into worldly activity.[134] But it was the Puritans who brought this concern with saving time to bear on worldly activity in an ascetic manner.

Closely bound to the Puritans' concern with time was the harsh attitude they held toward earthly pleasure. The Puritans were suspicious of pleasurable activity in general, claims Weber, and the worldly asceticism of the Puritans "turned with all its force against one thing: the spontaneous enjoyment of life and all it had to offer."[135] This tendency of Puritanism is exemplified in the following quote from Baxter, as is the close relation between this suspicion of pleasure and the concern with saving time:

> Keep up a high esteem of time, and be every day more careful that you lose none of your time, than you are that you lose none of your gold and silver. And if vain recreation, dressings, feastings, idle talk, unprofitable company, or sleep be any of them temptations to rob you of any of your time, accordingly heighten your watchfulness.[136]

This distrustful attitude toward earthly activities such as eating, dressing, and conversing is missing that counterbalancing element which created a tension in Calvin's teachings on the pleasures of this world. Calvin, who at some points insisted on the utter worthlessness of earthly life, at other points was able to claim that "there is not one little blade of grass, there is no color in the world that is not intended to make men rejoice." (See p. 128) It will be recalled that Calvin also argued that food was not simply a necessity, but also a source of "enjoyment and delight," and clothing was properly directed not toward necessity alone, but to "comeliness and honour" as well.

I will return to this difference between the Puritans and Calvin shortly, but first I must follow the last lines of Weber's argument. While this worldly asceticism of the Puritans can be interpreted as a safeguard against the dangers posed to righteousness by earthly success, this asceticism nonetheless complemented the beneficial effect the Puritans' unique conception of the calling had on capitalist development. As the Puritan entrepreneur restlessly devoted himself to his calling in order to prove his election to salvation, he was restrained from squandering his increasing wealth by Puritanical asceticism. This asceticism, claims Weber, "acted powerfully against the spontaneous enjoyment of possessions; it restricted consumption, especially of luxuries."[137]

At an early stage of capitalist development, such limitations of consumption helped to further that development. Weber notes that "the inevitable practical result" of the combination of the idea of the calling as proof and worldly asceticism, was the "accumulation of capital through ascetic compulsion to save."[138] These savings could then be invested in more productive activity, further increasing the glory of God.

But as was indicated in the earlier discussion of contemporary Marxist perspectives on consumption, capitalism would eventually require an augmentation of consumption in order to absorb excess productive capacity. (See discussion of Aglietta's notion of Fordism in Chapter 3, pp. 53–58.) Although Weber, who wrote *The Protestant Ethic* years before Henry Ford's production/consumption process began to roll, does not deal explicitly with capitalism's need to promote the consumption of commodities, he does acknowledge that a shift from limitation to augmentation of consumption had occurred in capitalism. In fact, the requirement to consume is a principal element of Weber's notion of the iron cage, which he introduced in the last few pages of his text. The following quote, in which the idea of the cage is first mentioned, highlights precisely this shift from the Puritanical restriction, to the modern promotion, of consumption:

> In Baxter's view the care for external goods should only lie on the shoulders of the 'saint like a light cloak, which can be thrown aside at any moment'. But fate decreed that the cloak should become an iron cage.
>
> Since asceticism undertook to remodel the world and to work out its ideals in the world, material goods have gained an increasing and finally an inexorable power over the lives of men as at no previous period in history.[139]

There is no doubt that, for Weber, the rationally organized capitalism of modernity is a cage in part because it offers no options but for people to take up their roles in this economic order. As Weber puts it, "The Puritan wanted to work in a calling; we are forced to do so."[140] But that is not all there is to the cage. The concern with "external" or "material" goods, the commodities which are produced by this economic order, is the feature which Weber explicitly mentions in his discussion of the cage. The Puritans' light cloak of material goods has become a cage for modern individuals.

Although Weber acknowledges that the religious asceticism of the Puritans, which limited consumption, "has escaped from the cage,"[141] he does not offer any explanation for this escape. He brings up this issue of the modern enslavement to commodities, perhaps, just to point up the difference between those conditions at the inception of the modern period, upon which his text focused, and the situation in which he was writing, early in the twentieth century. But I think that there is more to this difference than mere contrast, and that Weber brought up the issue of modern consumption and the cage at the end of his text in order to leave the reader with a question. And although he wrote that "fate decreed that the cloak should become an iron cage," I do not think Weber raised the question of how the cloak became a cage only to leave off his questioning with a reference to fate. That is, I do not think Weber raises this question in the pious, humble sense in which Augustine and Calvin raised the question of God's predestination of men and women, as a sort of sacred question mark beyond which one should not venture. Instead, as I read him, Weber raises the question to provoke one to think about it and attempt to answer it. I will attempt to do so in the next chapter.

CHAPTER 7

Nietzsche and Modern Asceticism

One must concede, at the outset, that any adequate explanation of the shift from the "worldly ascetic" limitation of consumption to modern fetishistic consumption must take into account the needs of capitalism itself, as indicated by those Marxists I examined in Chapter 3. Weber would most likely have been willing to make such a concession. In the final paragraph of *The Protestant Ethic*, Weber states that "it would also further be necessary to investigate how Protestant asceticism was in turn influenced in its development and its character [and, I might add, its dissolution] by the totality of social conditions, especially economic."[1] But the line of inquiry I want to follow in regard to the shift from limited to frenzied consumption is the one Weber "traced" in his text, the one having to do with religious ideas. I think one can tease an answer to this question about consumption out of Weber's argument itself, when Weber's argument about Protestantism is viewed from a different, and somewhat broader, perspective.

The broader perspective I have in mind here is that of Nietzsche, who was concerned not so much with the rise of capitalism, as was Weber, but with the history of nihilism. For Nietzsche, it was not so much the unanticipated, dynamic consequences of Protestantism which were important, as it was the reactive nature of Protestantism in the context of Christianity's progressive decline. In a note written in 1887, Nietzsche describes Protestantism as:

> that spiritually unclean and boring form of decadence in which Christianity has been able so far to preserve itself in the mediocre north; valuable for knowledge as something complex and a halfway house, in so far as it brought together in the same heads experiences of different orders and origins.[2]

This idea of Protestantism as a halfway house on the decline of Christianity is quite compatible with Weber's interpretation of Protestantism, especially in regard to Weber's emphasis on the importance of the doctrine of predestination. As I tried to emphasize in my elaboration of Weber's treatment of predestination, Calvin's resurrection of this doctrine in the troubling, disturbing duality in which Augustine had framed it was a challenge to the confidence and complacency which were spawned by the sacramental Church. Salvation was not only thrown into doubt, but was explicitly denied to the majority. From Nietzsche's perspective, Calvin's challenge to the Catholic Church's longstanding rejection of the double decree can be interpreted as an attempt to halt the decline of Christianity, to revive that religious intensity which accompanies the anxious concern with salvation. Calvin's doctrine of predestination brought the issue of the afterlife into sharp focus and held it before Christians and in this way could aid in rekindling religious fervor.[3]

Luther and Calvin's ambivalence toward earthly life, which I discussed earlier at some length (but which Weber himself does not develop), can also be interpreted in terms of the halfway house. By urging believers to both enjoy life and despise it, to see in earthly life both God's creative majesty and his righteous chastisement, both of these reformers can be seen as trying to heighten that tension within the Christian which had been weakened by Catholicism's increasing tolerance of worldly activity. By attempting to intensify the Christian ambivalence concerning life on earth, Luther and Calvin were trying to make the Christian a more responsive religious instrument, one closely tuned to its involvement with the things of this world. I see this heightened ambivalence of Luther and Calvin as a complement to the doctrine of predestination in the attempt to revive Christianity.

But what about the worldly asceticism of the Puritans, which on Weber's account played an important role in the development of modern capitalism? Does Nietzsche's perspective on Protestantism have much to offer on this particular feature? The initial answer to these questions would have to be "not explicitly;" Nietzsche does not specifically address Calvinist asceticism. But a more elaborate, interesting answer to these questions can be provided. To do this, it will be necessary to draw out Nietzsche even further, and to expound at some length what appears to be nothing more than an incidental remark of Nietzsche's.

In one particular note, in which Nietzsche criticizes German (i.e., Lutheran) Protestantism as being stale, lazy, and comfortably relaxed, he says of Protestantism, "A homoeopathy of Christianity is what I call it."[4] Homoeopathy was a nineteenth-century medical practice in which small doses of a poisonous drug were administered as a remedy for the sick. If administered to the healthy, these drugs would produce the same symptoms as those found in the sick.[5]

Although the note in which this remark about homoeopathy occurs appears to be quite incompatible with any discussion of Protestant asceticism (since the note emphasizes the laziness and comfortableness of Protestantism, and plays on the fact that homoeopathy uses weak doses of drugs), I am going to argue that this idea of Protestantism as homoeopathy is nevertheless helpful in understanding worldly asceticism. One can get around the initial obstacle that the context of Nietzsche's remark poses to any discussion of asceticism by pointing out that Nietzsche was referring in that note to German Lutheranism, not the Calvinist Puritanism that Weber stressed. But the notion of homeopathy can be extended beyond Nietzsche's discussion of Lutheranism. Although he does not use this particular term to describe it, Nietzsche's account of the activity of the "ascetic priest" in *The Genealogy of Morals* provides a clear example of homoeopathy, as well as a broader view of the worldly asceticism which Weber identified.

In the third essay of the *Genealogy*, Nietzsche emphasizes the importance of asceticism for the priest: it is "the main instrument of priestcraft, the supreme guarantee of their power."[6] Asceticism serves as a guarantee of priestly power, in that priests are the best examples of those who live according to the ascetic ideals they espouse. "The ascetic priest," claims Nietzsche, "is an incarnation of the wish to be different,"[7] and it is the difference which such priests attain through their asceticism that gives them their power. But asceticism is also the "main instrument of priestcraft," and by this Nietzsche means that ascetic ideals are employed by the priest in his ministerial activity. It is here that the connection between asceticism and homoeopathy becomes clear.

For Nietzsche, the priest serves a definite medicinal function. "We must look upon the ascetic priest as the predestined advocate and savior of a sick flock,"[8] writes Nietzsche. And although he often refers to the priest as a physician, ultimately Nietzsche rejects this description. "It is scarcely correct to call him a physician," Nietzsche says of the priest, "much as he likes to see himself venerated as a savior. What he combats is only the discomfort of the sufferer, not the cause of his suffering, not even the condition of illness itself."[9]

A more accurate description of the ascetic priest would seem to be that of a pharmacist or druggist. "To be sure," writes Nietzsche, "he carries with him balms and ointments."[10] And from Nietzsche's perspective, it was Christian priests who brought this pharmacological practice to its most highly developed form. Of Christianity, Nietzsche claims, "Never have so many restoratives, palliatives, narcotics been gathered together in one place."[11]

But these drugs can never result in a cure, because they are poisons. As Nietzsche puts it, "even as he [the priest] alleviates the pain of his patients he pours poison into their wounds."[12] Furthermore, in order to practice his special skill, the priest "must first create patients,"[13] and he does so by prescribing his poisonous drugs to the otherwise healthy. So the homeopathic activity of the priest is not focused solely on the sick; he also strives to gain the healthy as clients. For these reasons, Nietzsche writes that "wherever the ascetic priest has been able to enforce his treatment, the sickness has increased alarmingly, both in breadth and depth."[14]

In the *Genealogy* Nietzsche lists several of the "medications" which are used by the homeopathic priests to create and treat their patients. Despite the wide variety Nietzsche finds in the priestly "cabinet of hypnotic drugs,"[15] I would like to emphasize here only that the ascetic ideals of the priest are an important part of this pharmacy. Poverty, humility, and chastity, which Nietzsche identifies in the *Genealogy* as the "three mighty slogans of the ascetic ideal,"[16] are described in an earlier note as "dangerous and slanderous ideals" and as "poisons."[17] But Nietzsche recognizes the homeopathic benefits of these ascetic ideals, and that "in the case of certain illnesses" these poisonous ideals can be "indispensable as temporary cures."[18] So Nietzsche was not opposed to the homeopathic use of asceticism in general;[19] it was only the way in which the priests used it that bothered him.

According to Nietzsche, asceticism is one of the many things that was "ruined by the church's misuse of it."[20] For the Christian priests, asceticism was not a "temporary cure"; it was a way of life. And this way of life was based on a contempt for the body, the life processes, the sensuous. In contrast to Arendt (but perhaps as one-sidedly), Nietzsche found that Christians "despised the body; they left it out of account: more, they treated it as an enemy."[21] Christians were characterized by a "contempt for, and a deliberate desire to disregard the demands of the body."[22] For them, "suffering, struggle, work, death are considered as objections and question marks against life, as something that ought not to last; for which one requires a cure."[23] That cure, of course, is Christian asceticism.

To turn now to that specifically Protestant asceticism which so interested Weber, it should be obvious that Nietzsche could not have held it in very high esteem. And since Nietzsche did not have much to say about Protestantism generally, much less about Calvinist Protestantism, it is tempting to simply treat worldly asceticism as just another weak, dilute feature of this "spiritually unclean and boring" form of Christianity. But since the purpose of this examination of Nietzsche's perspective is to see whether he can help to answer the question of whether there is any relation between the limited consumption of ascetic Protestantism and modern consumption, I must spend a little more time here. As Nietzsche put it, Protestantism was "valuable for knowledge as something complex."

To begin with, it is not clear that Nietzsche would have had nothing but contempt for Protestant asceticism, at least the asceticism of Luther and Calvin. In fact, I think Nietzsche may have appreciated precisely that ambivalence which I tried to identify in Luther's and Calvin's writings. In the *Genealogy*, Nietzsche discusses the ascetic ideal of chastity (which I, for reasons already mentioned, have not discussed—see Chapter 6, note 62), and displays an appreciation for ambivalence. In this discussion, Nietzsche cites Luther's attitude toward chastity with approval, because it was not a one-sided adoration of chastity. "Perhaps Luther's greatest merit," writes Nietzsche, "was to have the courage of his sensuality."[24] For Nietzsche, "there is no inherent contradiction between chastity and sensual pleasure: every good marriage, every real love affair transcends these opposites."[25] According to the context in which these remarks were made, Nietzsche seemed to believe that Luther had such a marriage.

Now when Nietzsche says that these opposites are transcended, he is not referring to the dialectician's sense of transcendence as *aufheben*, where one of the opposed entities is advanced or elevated, in a new, superior form, and the other is left behind. As I read him, when Nietzsche mentions the transcendence of the opposites of chastity and sensuality, he means the transcendence of these things as mutually exclusive opposites. Nietzsche continues his discussion of chastity as follows:

> But even in cases where a real conflict exists between the sexual urge and chastity, the issue, fortunately, need not be tragic. At least this holds for all those happy, soundly constituted mortals who are far from regarding their precarious balance between beast and angel as an argument against existence. The finest and most luminous among them...have even seen in this conflict one more enticement to life.[26]

While Nietzsche most certainly would not have included Luther or Calvin among the brightest lights of the soundly constituted set (their ambivalence about earthly life was maintained with an eye constantly toward the afterlife, after all), I would like to assert that he nevertheless would have appreciated their ambivalence. The invigorating, if ultimately misguided, effects of their taut, tense ambivalence would not have been lost on Nietzsche. But the worldly asceticism of the Puritans, which smothered the celebratory, joyous dimension of Calvin's and Luther's ambivalence, is another matter.

After pointing out how the happy, well-constituted types could turn the conflict between chastity and sensuality into an enhancement of, or an enticement to, life, Nietzsche describes how the less appealing sorts would approach this virtue. "On the other hand, it is obvious that, once those pigs who have failed as pigs...come round to the worship of chastity, they will view it simply as their own opposite and will worship it with the most tragic grunting zeal."[27] I think Nietzsche would have had a similar judgment of the severe asceticism of the Puritans. Their suspicion of all pleasurable activities which might draw from the time spent glorifying God would surely have rankled Nietzsche, and in their time-saving slogans Nietzsche would likely have heard the squeal of swine. For these ascetics were so caught up in their worldly activity, in their callings, that they had to idolize that attitude which held life on earth to be worthless and despicable. In this way they could go about their earthly activity with a good conscience. Recall that even in Weber's scheme, worldly asceticism emerged as a ministerial tool, or drug, which was supposed to minimize the dangers posed by the idea of the calling as proof.

Now the fact that the asceticism of the Puritans promoted worldly activity must not be imagined to be to its credit when viewed from Nietzsche's perspective. Although Nietzsche complained bitterly of the otherworldliness of traditional priestly asceticism, the dogged, dreary pursuit of one's calling was not an alternative to otherworldliness which Nietzsche would have applauded. In fact, in the list of drugs he found in the priests' medicine cabinet, Nietzsche mentions "mechanical activity," which is quite similar to the Puritan's calling. "Mechanical activity, with its numerous implications (regular performance, punctual and automatic obedience, unvarying routine, a sanctioning, even an enjoining of impersonality, self-oblivion)—how thoroughly and subtly has the ascetic priest made use of it in his battle against pain!"[28] (It will be recalled that Luther and Calvin were pioneers in this disciplinary use of mechanical activity, or the calling. See Chapter 6, pp. 117–123.)

So, from Nietzsche's position, there is not very much new with worldly asceticism. The old ascetic slogan of poverty may have been abandoned by this new form of asceticism, and it may have, following Luther and Calvin, reversed the Pauline indifference to worldly activity, but the increased dosage of mechanical activity, in the form of the calling, made up for any decreases in those other medications. In the end, the patient was as sick as ever.

Therefore, Nietzsche's stance toward Protestant asceticism is, as one should expect, a complex one. While he may have approved of the ambivalence of Luther and Calvin, he would have rejected the worldly asceticism of the Calvinists. But the most important question for the purposes of this text remains to be asked. That question is whether, from a Nietzschean perspective, there is any connection between the worldly asceticism of the Puritans and the cage of modern consumption which Weber identified.

To answer this question, it must first be pointed out that for Nietzsche, asceticism did not end with Christianity. On the contrary, Nietzsche recognized in the most unlikely of places a new, particularly modern form of asceticism. He found this asceticism in the objectivity of scientific scholarship, which is frequently offered as the very opposite of religious asceticism. "People say to me that such a counterideal [to asceticism] exists," writes Nietzsche, "that not only has it waged a long, successful battle against asceticism but to all intents and purposes triumphed over it. The whole body of modern scholarship is cited in support of this."[29]

For Nietzsche, however, modern, scientific scholarship is not a foe of asceticism; rather, it is "in fact, its noblest and latest form."[30] It is the commitment of these scholars to the ideal of truth which gives them away as ascetics. For in their pursuit of scientific truth, modern scholars adopt their own unique ascetic regime: "it is necessary that the emotions be cooled, the tempo slowed down, that dialectic be put in place of instinct, that seriousness set its face on stamp and gesture."[31] And as for the "absolute will to truth" which drives such scholarship, Nietzsche charges that "it is nothing other than a belief in the ascetic ideal in its most radical form."[32]

As I stated at the very beginning of this examination of technology, however, my concern is not with science and the values that underlie it. I am concerned, instead, with the value that underlies the consumption of technology, a value I have identified as convenience. What I would now like to suggest is that this value of convenience, like the value or ideal of scientific truth, can be shown to be the basis for another late form of asceticism. In other words, my claim is that techno-fetishism is a form of asceticism. Furthermore, this modern form of asceticism can be shown to be related to the worldly asceticism Weber examined.

Now I am sure that this claim must appear even more implausible than Nietzsche's unusual interpretation of science as asceticism—at least scientists reject the testimony of their senses in their search for truth. The modern consumer of technology, however, would appear to be a slave to his or her senses and pleasurable sensations. Fast food, constant audio and visual entertainment, comfortable and speedy travel (and all of these comforts in a wide variety to choose from) would hardly seem to indicate an ascetic lifestyle. On the contrary, modernity would appear to be characterized by the constant titillation of the senses, the maximization of pleasure, the refusal to deny anything to the self. How can anyone possibly interpret this age of bliss as one that contains any trace of asceticism?

My response to this question would be to ask some different ones. How can one imagine that asceticism, which had been practiced and perfected for millennia by various priests, was overcome completely and permanently with the eclipse of religious belief, or the death of God, to use Nietzsche's notorious phrase? Can one really accept that the tremendous self-loathing which Nietzsche uncovered at the origin of Christianity, has been extinguished by that God's demise? Anyone who is at all receptive to Nietzsche's sensitive, perhaps hypersensitive, examination of asceticism would seem to have trouble imagining that modernity had gotten over this particularly human sickness. But even so, there is still quite a big step to be taken to get from the suspicion that asceticism must still be lurking in modernity to the conclusion that the consumption of modern technology is one of the forms which it has assumed. Weber's argument about worldly asceticism, when read through Nietzschean lenses, helps bridge this gap.

As I have already mentioned, Calvin's insistence on predestination as a double decree and his and Luther's ambivalence toward earthly life can be interpreted as efforts to preserve Christianity. But these efforts of the reformers were bound to fail, because their medicine was too strong and their patients were too sick to ever achieve a cure. In fact, new medications were required to counter the deleterious effects of those harsher drugs. The calling as proof of salvation and the severe attitude toward earthly pleasures were antidotes used by Puritan ministers to counter the prescriptions of Calvin.

Worldly asceticism, therefore, like the asceticism of earlier priests, was homeopathic medicine—the application of poisons to sick people—but it was, despite appearances to the contrary, a weaker form of medication than its predecessors. It allowed Christians to throw themselves without compunction into worldly activity. Charitable works were no longer the hallmark of Christian activity; successful business enterprises became the sign of God's presence in the world. Even though the Puritans limited the enjoyment of earthly activities and things, their asceticism provided proof of immortality, of the eventual relief from the toils and troubles of mortal existence.

But even this weaker form of asceticism could not stop the progress of the nihilistic disease. Faith in God, or in the possibility of immortality, which was both the premise and promise of the Christian God, eventually became untenable. This is not to say that people are no longer willing to profess their faith, either by celebrating rituals with other believers in some church congregation or by sending checks to the saints of the television satellites. What I am saying is that, in the most developed of those cultures in which Christianity flourished, earthly activity, by and large, is no longer undertaken with a view to heavenly immortality. The idea of heavenly immortality has receded from the forefront of the modern horizon. It is no longer a guiding principle of human actions on, and beyond, earth. For now I am going to simply assert this claim about the demise of Christianity, which is sure to be rejected by some, although evidence in support of this claim, if not proof, could be offered. My argument presupposes that this claim will not present a stumbling block to most readers.

To continue this line of assertions, I want to further claim that, even though God may have receded or retreated from the modern world, the need for a God remains. Modern individuals have not become well, in the Nietzschean sense that they celebrate their mortality, their embodiment, their senses, both pleasurable and painful. All of these conditions still remain a source of anxiety to humans, but the projection of this anxiety into a supersensuous realm of immortality, access to which is determined by God, will no longer suffice to comfort most moderns. A new drug is needed; a new ascetic practice is required. My claim is that convenience is that drug, and the consumption of technology is that practice.

As I argued earlier when I challenged Arendt's interpretation of modernity, the tremendous productive capacity of modernity and the heightened concern for maintaining and increasing that capacity are not indications of any modern "reverence for the body." On the contrary, if one takes into account what it is that is consumed by modern producers, it becomes apparent that modernity is characterized by a certain revulsion against the body, mortality, and necessity. The demands of the body, which were ignored or strictly regulated by Christian asceticism, in both its monastic and Puritanical forms, are no longer something to be neglected or restricted. Instead, they have become, to recall the distinction I introduced earlier, limits imposed by the body. And the overcoming of these limits is the value of convenience, in the particularly modern sense of this word.

The asceticism which I am trying to identify may appear at first glance to be the opposite of its Puritanical predecessor. The latter restricted consumption, while modern asceticism, as I have stretched the term, is based on continually increasing consumption. But there are certain similarities between worldly asceticism and modern techno-fetishism, aside from their relation to human mortality.

The Puritans' concern for saving time, for not wasting a moment, is also present, although in an altered form, in modern asceticism. For the Puritan, any time spent outside the performance of one's calling was, strictly speaking, wasted time. For moderns, it is not time spent away from the calling which must be minimized, but time spent in the satisfaction of the demands of the body. I have already discussed as examples of this attitude toward time certain developments in the production and preparation of food and the means of transportation (although I earlier emphasized the material conditions of the United States as a factor in that modern attitude). Here, let me just mention that the body's demands for clothing (e.g., shopping by phone or mail on credit) and shelter (e.g., prefabricated homes, maintenance-free condominiums) have also become satisfied much more quickly in modernity than ever before. Along with the material and economic conditions which played a role in this 'saving' of time, there is also this element of modern asceticism.

Just as there is no proof available to convince modern Christians that God has retreated from the world, there is no way to prove to the techno-fetishist that modern consumption practices have anything to do with the death of God. But here, too, evidence is available. Such evidence, drawn from modern political thought, will be presented in the following chapter of this text. But first I want to examine a particularly ironic bit of evidence which not only supports my claim that modern asceticism is related to the death of God, but does so by extending and expanding Weber's argument concerning worldly asceticism.

Early in *The Protestant Ethic*, Weber offers the writings of Benjamin Franklin as "a document of that [capitalist] spirit which contains what we are looking for in almost classical purity."[33] In Franklin's books of ethical maxims, Weber finds examples of the worldly asceticism of Puritanism, without that religious context. For Weber, Franklin is on the cusp; he represents the transition from the age in which a calling was pursued for religious reasons to the age of the cage, in which a calling is pursued for utilitarian reasons, if indeed, there is any choice involved at all.

For Franklin, the successful pursuit of a calling was not undertaken to prove one's predestination to salvation,[34] nor was wealth accumulated in order to provide for a life of leisure and comfort. According to Weber, Franklin valued the accumulation of wealth in itself, and not for what it could prove or provide. Weber describes the transitional character of Franklin's ethic as follows:

> The earning of more and more money, combined with the strict avoidance of all spontaneous enjoyment of life, is above all completely devoid of any eudaemonistic, not to say hedonistic, admixture. It is thought of so purely as an end in itself, that from the point of view of the happiness of, or utility to, the single individual, it appears entirely transcendental and absolutely irrational. Man is dominated by the making of money, by acquisition as the ultimate purpose of his life.[35]

Now even though Franklin may have abandoned the Calvinism of his parents, his advice on how to accumulate wealth sounded somewhat like those Calvinist ministers such as Baxter, whom Weber cited to elucidate Puritan asceticism. In Franklin's *Necessary Hints to Those That Would be Rich* (1736) and *Advice to a Young Tradesman* (1748), which Weber quotes at length in his text, there is that concern with time which characterized Puritan asceticism, but for Franklin time was valuable not because it could be spent glorifying God through one's calling. Instead, Franklin claimed "that *time* is money. He that can earn ten shillings a day by his labour, and goes abroad, or sits idle, one half of that day...has really spent, or rather thrown away, five shillings...."[36] In a later, widely read essay entitled "The Way to Wealth," Franklin is even more emphatic about the value of time.

> But *doest thou love Life, then do not squander Time, for that's the Stuff Life is made of,* as *Poor Richard* says.... If Time be of all Things the most precious, *wasting Time* must be, as *Poor Richard* says, *the greatest prodigality.*[37]

In this same essay, Franklin also discusses that other dimension of Protestant asceticism, restricted consumption. Although Weber does not cite this essay and does not really discuss Franklin's attitude toward consumption, a particular passage from it supports Weber's interpretation of Franklin as a sort of nonreligious ascetic and also serves my argument about modern asceticism.

> Here you are all got together at this Vendue of *Fineries* and *Knicknacks*. You call them *Goods*, but if you do not take Care, they will prove *Evils* to some of you. You expect they will be sold *cheap*, and perhaps they may for less than they cost; but if you have no Occasion for them, they must be *dear* to you. Remember what *Poor Richard* says, *Buy what thou hast no Need of, and ere long thou shalt sell thy Necessaries.*[38]

In this warning against excessive consumption, Franklin employs that necessary/artificial distinction which has surfaced in several places and forms throughout this genealogy of convenience. Of the non-essential goods mentioned in the preceding quote, Franklin continues, "These are not the *Necessaries* of Life; they can scarcely be called the *Conveniencies*, and yet only because they look pretty, how many *want* to *have* them. The *artificial* Wants of Mankind thus become more numerous than the natural."[39] Franklin recognized in this increase of wants a certain danger, as did the Calvinist ministers, and he warns against this trend. He is especially suspicious of the purchase of such goods or commodities on terms of credit. "But what Madness must it be to *run in Debt* for these Superfluities!"[40] warns Franklin.

So in both his concern with wasted time and his restrictive attitude toward consumption, Franklin does appear to be advocating a form of asceticism which closely resembles the worldly asceticism of Puritanism. But I want to suggest that Franklin was not simply a Puritan stripped of his religious foundation, as Weber seems to suggest. Franklin was indeed ascetic, but he displays elements of the modern asceticism I am trying to bring to light. (Franklin's disdain for consumption on credit is not one of these modern elements.)

A first glimpse of this asceticism can be found in the above quote concerning necessity and artificiality. Franklin's use of this distinction differs from that of earlier thinkers such as Augustine and Luther because he introduces a new category of earthly possession. Franklin seems to be saying that if such goods were definitely among the "Conveniencies," then they would be less artificial or more necessary and would therefore be less troubling or dangerous. Franklin's notion of necessity seems to be expanding to include the need for convenience.

This impression from the quote is further supported by the fact that about midway through his life, at the age of forty-two, Franklin appears to have undergone a shift in his attitude toward time. At this point, Franklin retired from his printing business, although he still received a share of the profit from that business for many years afterwards.[41] But Franklin no longer spent his time actively pursuing wealth, as he had advised others to do. In a letter to a friend, written in the year in which he retired, Franklin describes his new life:

> Thus you see I am in a fair way of having no other tasks, than such as I shall like to give myself, and of enjoying what I look upon as a great happiness, leisure to read, study, make experiments, and converse at large with such ingenious and worthy men, as are pleased to honour me with their friendship or acquaintance, on such points as may produce something for the common benefit of mankind.[42]

Franklin even urged friends to follow his example. In another letter to another friend, he asked,

> By the way, when do you intend to live—i.e., to enjoy life...will you retire to your villa, give yourself repose, delight

> in viewing the operations of nature in the vegetable creation, assist her in her works, get your ingenious friends at times about you, make them happy with your conversation, and enjoy theirs: or, if alone, amuse yourself with your books and elegant collections.[43]

This advice concerning the economy of time is certainly quite different from that offered by either Richard Baxter, Weber's prototypical Calvinist, or Poor Richard, Weber's model of the ascetic entrepreneur. Both of these ascetics would have looked upon the advice of the retired, leisurely Franklin as an invitation to waste time. Retire to your villa? Give yourself repose? Amuse yourself with books and collections? Such activity, or lack of activity, would not have augmented the glory of God, according to Baxter, nor would it have augmented one's wealth, as Poor Richard says. But for the mature Franklin, such leisurely pursuits were not a waste of time because they were all directed toward the "common benefit of mankind."

For Franklin, humanity could be served not just by the successful performance of an earthly calling, such as his business as a printer prior to retirement. Humans could also benefit from the scientific investigation of the laws of nature, and the application of those laws to mortal, earthly conditions. Franklin valued the time he spent performing experiments and designing 'improvements' for humanity and retired so that he would have more of such time. Even though his retirement was disrupted by his full and varied career in public service, Franklin nevertheless became a prominent figure in eighteenth-century 'natural philosophy,' and designed several important devices over the course of his life. A brief examination of some of these devices will make it clearer what Franklin had in mind when he wrote about the common benefit of humanity.

Borrowing from German designs, Franklin developed a woodburning stove that became enormously popular. The advantages of the 'Pennsylvania fireplace' (or Franklin stove, as it came to be known) were many, according to a pamphlet Franklin wrote to promote its sale.[44] But the principal benefit of the stove was that it was much more efficient than an open fire. Franklin's design made use of the hot gases which, in a common fireplace, rise directly into the chimney. In Franklin's stove, those gases were used to heat a thick metal plate, which in turn heated the air above and around it. The result was "that your whole Room is equally warmed; so that People need not croud so close round the Fire, but may sit near the Window, and have the Benefit of the Light for Reading, Writing, Needlework, &c. They may sit with Comfort in any Part of the Room, which is a very considerable Advantage in a large Family, where there must often be two Fires kept, because all cannot conveniently come at one."[45]

In other words, the stove freed people from the hearth and allowed them to go about other activities in the heated room. Staying warm was no longer as much of a burden, and the stove saved some of the time that had been taken up by the body's demand for heat. And since these stoves were more efficient and used less wood than fireplaces, they also shortened the amount of time that one had to spend moving wood about to feed the fire. The stoves were also easier to light and safer than open fires. In conclusion, Franklin says of the stove, "With all these Conveniences, you do not lose the pleasing Sight nor Use of the Fire, as in the Dutch Stoves, but may boil the Tea-Kettle, warm the Flat-Irons, heat Heaters, keep warm a Dish of Victuals by setting it on the Top, &c. &c."[46] So besides heating bodies, the stove could simultaneously be used to help perform other household tasks. It is in this sense of improving efficiency in the necessary activity of the household, of speeding things up, that the Franklin stove can be described as a convenience.

Franklin, as is well known, was also a pioneer in the study of electricity, and here too he put this knowledge in the service of convenience. Aside from his invention of the lightning rod, Franklin also used his knowledge of electricity to kill animals to be eaten. One advantage of electrocution was that it resulted in immediate, sudden death and was therefore thought to be more humane than other methods. But electrocution also helped to minimize the time that meat had to hang in order to become tender. "The flesh of animals, fresh-killed in the usual manner, is firm, hard and not in a very eatable state..." wrote Franklin to some friends, but "in its progress towards putrefaction...the flesh becomes what we call tender, or is in that state most proper to be used as our food."[47] But when the animal is electrocuted in the manner which Franklin fully described, "the putrefaction sometimes proceeds with surprising celerity."[48] For Franklin, this acceleration of "putrefaction" was the chief recommendation for electric slaughtering.

It is also interesting to note that Franklin was a leader in making the production and storage of electricity more convenient in itself. The first experiments with electricity were performed with the "Leyden jar," a glass tube rubbed with a piece of silk, which thereby condensed the electric charge. Soon after Franklin received such a tube from a friend in England, he wrote back to that friend:

> The *European* papers on Electricity, frequently speak of rubbing the tube, as a fatiguing exercise. Our spheres are fixed on iron axes, which pass through them. At one end of the axis there is a small handle, with which you turn the sphere like a common grindstone. This we find very commodious, as the machine takes up but little room, is portable, and may be enclosed in a tight box, when not in use.[49]

Franklin and his American colleagues, therefore, not only performed experiments along the lines which had been established in Europe; they also facilitated the performance of such experiments by creating portable generators.

Franklin even turned his invention of the lightning rod into a device for speeding up the investigation of electricity, and he did so in a manner that indicates the direction in which the development of electricity would be carried out. One of the lightning rods on Franklin's house in Philadelphia ran not to the ground outside the house, to render lightning harmless, but instead into the house itself and then to a ground. Franklin attached two bells to the wire running through his house, and these sounded whenever an electrical charge was being drawn through the lightning rod. In a letter describing this arrangement, Franklin claimed to have "frequently drawn sparks and charged bottles" from this device.[50] Through this technique, Franklin eliminated the necessity of rubbing the bottle or turning the crank; all he had to do to acquire a charge for his experiments was draw off electricity generated in the atmosphere. And Franklin's device even notified him when it was time to charge his bottles, so that he did not have to spend time waiting for the proper moment.

I would like to mention one last example of Franklin's inventiveness, an example that comes from the realm of transportation technology. Upon his return to the United States after many years as the American ambassador to France, Franklin designed a sedan chair which was used to move him about the city of Philadelphia. There is nothing truly innovative here—such conveyances had been developed over a century before in Europe—but there is something about Franklin's use of the sedan chair that illuminates the point I am trying to make. Early in the seventeenth century, John Winthrop, the Puritan governor of Massachusetts Bay Colony, had refused to accept the gift of a sedan chair, and even toward the end of that century, horse-drawn carriages were frowned upon in Boston as things of this world only.[51] Roughly a century later, however, Franklin had no religious compunction about such worldly things as sedan chairs. Indeed, Franklin wrote that he wished "I had brought with me from France a balloon sufficiently large to raise me from the ground. In my malady it would have been the most easy carriage for me, being led by a string held by a man walking on the ground."[52] Such a wish would most likely have been worthy of punishment in seventeenth-century New England, but it was appropriate and prescient in eighteenth-century Philadelphia.

My point in discussing Franklin is not to contradict Weber's interpretation of him as an entrepreneurial ascetic. My claim is not that Franklin was, contra Weber, a libertine, or a lover of luxury. Those devices Franklin invented and developed were not, in his eyes, "superfluities" of the sort he warned against in his books of ethical maxims. If these things were not absolute necessities, they were "conveniencies," and as such they were valid and valuable.

I do want to point out, however, that Franklin was not as free from religion as Weber's portrayal of him implies. Although Franklin claimed to have abandoned religious disputation early in his career,[53] toward the end of his life he did set down his religious beliefs in a letter to Ezra Stiles, the President of Yale College. In that letter, written a month before his death, Franklin claimed "that the most acceptable Service we render to him [i.e., God] is doing good to his other Children," and "that the soul of Man is immortal, and will be treated with Justice in another Life respecting its Conduct in this."[54]

These beliefs, based as they are on the idea that earthly works have some bearing on one's salvation, fly in the face of the doctrine of predestination and the reformers' denigration of good works. But what I want to stress is not the fact that Franklin completely abandoned the Calvinism of his parents. Rather, I want to emphasize that Franklin's inventiveness, his skill at applying scientific knowledge to make life on earth more convenient and comfortable, had a religious sanction. In providing, for the "common benefit of mankind," devices such as efficient stoves and lightning rods, Franklin was not concerned with making money, as one might expect given Weber's argument. In fact, Franklin refused to accept patents on his inventions.[55] Instead, Franklin's motives were otherworldly; his concern was with the salvation of his soul, not earthly treasures.

From Weber's perspective, Benjamin Franklin was a transitional figure. He represents to Weber the worldly asceticism of the Puritans without that religious foundation. And for Weber, the worldly asceticism of Franklin was eventually eclipsed in the modern [c]age. From my perspective as well, Franklin is on the cusp, but for me he represents a new form of asceticism. And as I interpret modernity, even though the religious concern with an afterlife may have waned, the new form of asceticism has flourished as more and more time has been saved from bodily necessity.

CHAPTER 8

Traces of Modern Asceticism

HOBBES AND MORTALITY

While Franklin may have provided an ascetic example of "almost classical purity" to Weber, he is not quite as helpful in regard to the asceticism I am trying to identify. Modern asceticism, which is bound up with the fetishistic consumption of technology, is not supported by religious aspirations such as those voiced by Franklin late in his life. Specifically, modern asceticism is not ultimately grounded in the goal of an otherworldly immortality, as was Christian asceticism in its various forms. The celibacy and regularity of monastic orders and the worldly asceticism of the Puritans were directed toward that which was promised by Christ—life everlasting. And for his part, Franklin thought that the benefits he provided to everyone in common would gain for him this prize.

Modern asceticism, on the other hand, is not grounded in the Christian idea of an afterlife. While this asceticism has indeed come to embrace the promise of immortality, it is not an otherworldly afterlife toward which this asceticism aims. I will discuss this recent type of immortality in the following chapter. What I want to stress at this point is that modern asceticism, even though it was not able to abandon its Christian framework all at once, is distinguished from Christian asceticism partly but precisely by its rejection, tacit or otherwise, of Christian immortality. This rejection of Christian immortality, I will argue, is closely bound up with that other distinguishing feature of modern asceticism, the consumption of convenience.

To support these claims and to begin establishing in more detail the dimensions of modern asceticism, I will have to move beyond the limited example of Franklin and his inventions and examine other modern thinkers whose connection with technological development is not as clear as Franklin's. However, these other thinkers—Hobbes, Locke, Marx, and Marcuse—have the advantage of highlighting the relation between technological development and the mortal, finite body. It is in their non-Christian attitudes toward mortality and necessity that these four can be interpreted as modern ascetics. My choice of these theorists is based solely on their appropriateness to the argument I am making—which is, in its own way, narrow and focused. I do not want to give the impression that I have objectively surveyed modern political thought (or even these four thinkers, for that matter) and am offering a distillation of that experience. On the contrary, this reading is biased; it slashes the surface.

I should also offer at this point some explanation for the direction my argument has taken. It certainly may be asked why I have chosen to focus on political theorists in this penultimate chapter. Why not continue examining technological apparatuses such as those that Franklin invented? Could I not make my case for modern asceticism by referring to those devices that are consumed in order to deny the limits of the body? The answer to these questions is yes, the argument could be carried in that direction. And I imagine that it should be obvious by now how I would interpret contemporary technological developments such as satellite communications, space shuttles, organ transplants, test-tube babies, and so on. Developments such as these can all be interpreted as means for overcoming the temporal and spatial limits of embodiment.

But as I stated at the outset of this critique of technofetishism, my concern is not primarily with mapping the development and deployment of technology in modernity, but with uncovering or unearthing the value of convenience, roots and all. Given this genealogical objective, the four theorists listed above are more appropriate than a survey of recent technological developments. These theorists help to make the case that the current attitudes toward technology are based on a particularly modern attitude toward the body, mortality, and necessity.

I will pursue a chronological direction in this examination of modern thought, beginning with Thomas Hobbes, and I will focus on the mature version of Hobbes's account of the formation of civil society—*Leviathan*. As the subtitle of that text indicates, Hobbes, despite his reputation as one of the first theorists of the modern state, was not unconcerned with religious questions but was, on the contrary, concerned with "the Matter, Forme and Power of a Common-Wealth Ecclesiasticall and Civill."[1]

In my treatment of *Leviathan*, I will emphasize to some extent the religious dimension of Hobbes's thought, although I will avoid getting involved in the dichotomous argument concerning Hobbes's atheism or Christianity.[2] What is important for me is not the question of whether or not Hobbes was an atheist, but rather the way in which Hobbes introduces a new, non-Christian attitude toward mortality and necessity while remaining very much within the structure of Christian discourse. It is this attitude, not Hobbes's religious convictions, that I want to stress.

The issue of human mortality runs throughout Hobbes's argument in *Leviathan*, particularly in the state of nature he describes there. Due to two fundamental features of this state, the mortality of the human condition is never long out of mind. The first of these features is the natural right that every individual has "to use his own power, as he will himselfe, for the preservation of his own Nature; that is to say, of his own Life; and consequently, of doing any thing, which in his own Judgment, and Reason, hee shall conceive to be the aptest means thereunto."[3] Besides this natural right, Hobbes also recognizes in the state of nature a physical and mental equality among humans, upon which "ariseth equality of hope in the attaining of [their] Ends."[4]

Given these two features of the state of nature, when two individuals in this state of equality come to desire the same thing, their parity leads them to become competitors. Since neither of them has such an advantage that one can hope to intimidate the other into abandoning that common desire or forfeiting their natural right to that object of contention, they come to regard each other as enemies in a struggle to assert their natural right. This natural competition is intensified by the passions, primarily vanity. Hobbes describes this passion as follows:

> For every man looketh that his companion should value him, at the same rate he sets upon himselfe: And upon all signes of contempt, or undervaluing, naturally endeavors, as far as he dares, (which amongst them that have no power to keep them

> in quiet, is far enough to make them destroy each other), to extort a greater value from his contemners, by dommage [damage]; and from others, by the example.[5]

Leo Strauss has demonstrated at length how the competition that occurs in Hobbes's state of nature expands into a life and death struggle in the presence of the uncontrolled appetite, or passion, of vanity.[6] I need not set out Strauss's argument here, but can just say that he helps to explain Hobbes's conclusion that the state of nature is always in danger of turning into a state of war, "every man, against every man."[7] And as far as Hobbes is concerned, as long as the possibility of outright mortal conflict is imminent, there is a state of war.[8]

What I would like to emphasize about Hobbes's state of nature, however, is not its natural equality or the role that vanity plays in it, but the fact that it raises the issue of human mortality and does so in a temporal context.One of the problems or "incommodities" of the state of nature, to use Hobbes's term, is that peoples' lives in that state tend to be of short duration. In his famous description of life in this state of nature, Hobbes says it is "solitary, poor, nasty, brutish, and short."[9] At another point in the text, Hobbes says that in the state of natural equality, "there can be no security to any man, (how strong or wise soever he be,) of living out the time, which Nature ordinarily alloweth men to live."[10] In the state of nature, in other words, individuals are sold short, or shortchanged, on their time; they die not only violently, as Strauss emphasizes, but prematurely, as measured by natural standards.

There are, of course, natural forces or causes that lead people out of this state of nature and allow them to buy some time or extend the duration of their lives. The possibility for an individual to come out of the state of nature, Hobbes claims, consists "partly in the Passions, partly in his Reason."[11] It is human nature, therefore, to be inclined by the passions and reason to leave the state of nature. But like the state of nature, these passions and reason are also imbued with a sense of mortality and are guided by considerations of death.

One of the forces that leads out of the state of nature—reason—does so by placing a limitation on the exercise of the natural right to do and take whatever one wants. This limitation comes in the form of a law of nature. Hobbes stresses that a law of nature, like any law, is an obligation to do nor not to do something, whereas a right of nature, like any right, is a liberty to do or forego doing something.[12] And even though individuals in the state of nature tend to be rather nasty and brutish, they are reasonable enough, when not in a passionate frenzy, to recognize the laws of nature and accept some limitations on their natural rights.

In his discussion of the laws of nature, however, Hobbes defines a law of nature, in its generic sense, in a surprisingly specific manner.

> A LAW OF NATURE, (Lex Naturalis,) is a Precept or general Rule, found out by Reason, by which a man is forbidden to do, that, which is destructive of his life, or taketh away the means of preserving the same; and to omit, that, by which he thinketh it may be best preserved.[13]

A law of nature, therefore, is not just an obligation or limitation that humans naturally recognize; it is a limitation on the fundamental right to do what one thinks will best preserve one's life. According to the very idea of natural law, one cannot reasonably exercise one's natural right in a manner that would actually lead to one's demise. Furthermore, one must do, or as Hobbes put it, one is forbidden to omit doing, that which one thinks will best preserve or prolong one's life.

Hobbes gets even more specific about natural law after defining the category and names two such laws. The fundamental law of nature is "to seek Peace, and follow it."[14] The second such law, which is derived from the first, is "that a man be willing, when others are so too, as farre-forth, as for Peace, and defence of himselfe he shall think it necessary, to lay down this right to all things; and be contented with so much liberty against other men, as he would allow other men against himselfe."[15] So as Hobbes reads the laws of nature, human beings are naturally inclined, by dint of their reason, not to exercise their natural rights in a manner which could lead to violent competition and premature death. Rather, the individual is naturally inclined to sacrifice its rights, in common with others, in order to attain and preserve peace, but—most importantly—to save its mortal life.

This natural, reasonable inclination toward peace and civil society, of course, is just that, an inclination. Given Hobbes's view of the violent character of the state of nature, it is apparent that the laws of nature do not thoroughly determine human behavior, and that people can ignore those precepts that are discovered by their natural reason. For instance, an individual might be willing to put himself at risk in order to have revenge upon someone who slighted or insulted him. But there are other elements in the constitution of Hobbes's natural subject that come to the aid of reason and help to promote peace. These other elements are found among the passions, and foremost among them, in regard to the inclination toward peace, is the fear of violent death.[16]

On Hobbes's account, of all the incommodities of the natural state of war the "worst of all" is the "continual feare, and danger of violent death."[17] And it is this worst feature of the war which urges people most strongly toward peace. Here again, Leo Strauss's interpretation of Hobbes is instructive. After describing how the passion of vanity leads individuals in Hobbes's state of nature to engage in a physical struggle to have revenge on another for some real or imagined wrong, Strauss traces the development of this struggle to another level.

> At some point in the conflict, actual injury, or, more accurately, physical pain, arouses a fear for life. Fear moderates anger, puts the sense of being slighted into the background, and transforms the desire for revenge into hatred. The aim of the hater is no longer triumph over the enemy, but his death. The struggle for pre-eminence, about 'trifles', has become a life-and-death struggle. In this way natural man happens unforeseen upon the danger of death; in this way he comes to know this primary and greatest and supreme evil in the moment of being irresistibly driven to fall back before death in order to struggle for his life.[18]

It is this face-to-face confrontation with death, with the possibility that one's time may soon be over, that clarifies and intensifies the reasonableness of the laws of nature. The cold breath of death, one might say, compels the individual in the state of nature to seek peace in some form of civil society, where one no longer has to rely on oneself for protection. But while death is the goad that impels individuals toward civil peace, it is also the outer limit on the exercise of any civil authority, because it marks the spot beyond which the Hobbesian subject cannot go in its renunciation or transfer of its rights.

For Hobbes, "there be some Rights, which no man can be understood by any words, or other signes, to have abandoned, or transferred."[19] The first of these is "the right of resisting them, that assault him by force, to take away his life."[20] So the threat of death is at the heart, or throat, of the relation between the individual and the Hobbesian commonwealth. In those situations in which one's life is threatened, one need not obey any authority and can act out of one's most fundamental, natural right.

This is not to say that the sovereign authority cannot issue the punishment of death, however. In Hobbes's scheme, the civil authority of the commonwealth is the sole dispenser of punishments for breaches of particular laws or breaches of the very covenant upon which the commonwealth rests.[21] Among the punishments Hobbes lists in *Leviathan* are corporal punishments, "such as are stripes, or wounds, or deprivation of such pleasures of the body, as were before lawfully enjoyed,"[22] and he also includes capital punishment.[23]

But even though Hobbes's absolute sovereign wields the "power of life and death,"[24] it cannot compel an individual to kill or injure him or herself.

> If the Soveraign command a man, (though justly condemned,) to kill, wound, or mayme himself; or not to resist those that assault him; or to abstain from the use of food, ayre, medicine, or any other thing, without which he cannot live; yet hath that man the Liberty to disobey.[25]

In this statement, which is taken from Hobbes's description of the liberty of subjects, he not only recognizes the right to disobey orders to kill or injure oneself, but also implies the right to resist attempts by the sovereign and its agents to inflict punishments. Later in the text, in the chapter titled "Of Punishment and Rewards," Hobbes makes this explicit.

He begins this chapter by making the surprising concession that no one has to submit to punishment by the sovereign: "For by that which has been said before, no man is supposed bound by Covenant, not to resist violence; and consequently it cannot be intended that he gave any right to another to lay violent hands upon his person."[26] Hobbes argues that this natural, inalienable right to resist "is granted to be true by all men, in that they lead Criminals to Execution, and Prison, with armed men, notwithstanding that such Criminals have consented to the Law, by which they are condemned."[27]

So the limit of the subject's renunciation of its rights to the commonwealth, as well as the inclination to renounce anything in the first place, are derived from the same source—the mortal, finite body. It is the fear of death, of the end of one's time on earth, which crystallizes the reasonableness of the laws of nature for the natural subject, and it is this same fear which justifies resistance to the sovereign when punishment is threatened. For "a man cannot tell, when he seeth men [including policemen] proceed against him by violence, whether they intend his death or not."[28]

One should not overplay the role that death, or the fear of death, plays in Hobbes's thought, however. There are other passions which also lead people out of the state of nature, and these other passions may be grouped together as the desire for convenience, or "commodious living," to use Hobbes's term. Along with the fear of violent, premature death, Hobbes adds to the list of the passions which incline people toward peace the "Desire of such things as are necessary to commodious living; and a Hope by their Industry to obtain them."[29] This desire or need is one of the dimensions of the asceticism I am trying to identify in modernity.

So the role the body plays in Hobbes thought is not limited to the fear of death. In forming covenants to promote peace, Hobbes's subject is looking not just to lengthen its life, but to overcome those "nasty, brutish" aspects of life in the state of nature. Hobbes's conception of necessity, like Calvin's before him and Franklin's after, was not restricted to absolute physical necessity. Human needs included those things which made life commodious, or convenient, as Franklin would say.[30] Hobbes's claim that individuals join covenants in the hope that they will be able to obtain the commodities necessary for commodious living indicates that he was not under the sway of Puritan asceticism with its suspicious attitude toward worldly goods. But even though Hobbes was no Puritan, he does represent an early form of that modern asceticism which denies the body's limits through the consumption of convenience.

What I want to stress about Hobbes's thought, however, is not the status he affords to comfort and convenience. (I will focus on that dimension of modern asceticism when I get to Locke.) Rather, I would like to emphasize the less obvious connection between the role that human mortality and death play in Hobbes's thought and modern asceticism. As I mentioned at the beginning of this chapter, modern asceticism is distinguished from Christian asceticism by the stance it takes toward human mortality and the possibility of immortality. This difference can be brought out by a comparison of the worldly asceticism of the Puritans and the asceticism I have found in Hobbes's thought.

At first glance, there appears to be a certain similarity between these two forms of asceticism—they both can be interpreted as responses to anxiety concerning human mortality. Calvin made his followers anxiously aware of their mortality through his resurrection of the doctrine of gratuitous, double predestination, and the Puritans' conception of the calling and their asceticism were responses to this "decretum horribile" of Calvin's. Like Calvin, Hobbes also raises the issue of human mortality, but he does so through his description of the state of nature, where violent, premature death always threatens.

But when Hobbes raises the issue of mortality, at least in the first two parts of *Leviathan*, it is not in relation or regard to immortality. Death, in this part of Hobbes's argument, does not hold forth the possibility of eternal life. Rather, death for Hobbes was, as Leo Strauss put it, "the greatest and supreme evil." In my reading of Hobbes, death is evil because it is the ultimate temporal limit which is imposed on the subject by the body. The death with which the individual comes face-to-face in the life and death struggle reveals the essentially finite, temporal being of humans. If the subject dies, its time is up, so it must do whatever it can to avoid death and increase that time which nature allots to humans.

It is not just the desire for a commodious life, therefore, that fits Hobbes within the frame of my argument about convenience. The role that the fear of death plays in his thought can also be interpreted as an instance of the modern concern with overcoming the limits of the body—in this case, the ultimate temporal limit, which is death. This interpretation of Hobbes, however, would appear to hold, if at all, only for the first two parts, or the first half, of *Leviathan*. In the third part, "Of a Christian Common-Wealth," Hobbes does recognize the possibility that there is something beyond the limit of death, and even goes so far as to claim that if the civil sovereign issues a command which "cannot be obeyed, without [the subject] being damned to Eternall Death, then it were madnesse to obey it."[31]

This appears to contradict my claim that the life and death of the finite, mortal body presents in Hobbes's thought, the ultimate limit of sovereign authority as well as a limit to the renunciation of natural rights. For one could conceivably be commanded by the sovereign to do something upon the penalty of death which would result in one's banishment from the kingdom of God and the loss of one's eternal life. According to Hobbes, at least in this later part of *Leviathan*, one would be mad to abandon one's eternal life in such a situation simply to escape physical death; one should instead give up one's mortal life in order to save the immortal one. In this advice, Hobbes seems to recognize another temporal dimension beyond earthly time, one that is infinite or eternal.

But situations like the one described above, in which God's authority conflicts with that of a civil sovereign, should not really present as much trouble as they have in the past, writes Hobbes.[32] To avoid such conflicts, people "need to be taught to distinguish well between what is, and what is not Necessary to Eternall Salvation."[33] And in the chapter of *Leviathan* entitled "Of what is NECESSARY for a Mans Reception into the Kingdome of Heaven," Hobbes offers a lesson in making this distinction. But if Hobbes's lesson is followed closely, one can see that Hobbes really does not contradict my earlier claims about the mortal body, but instead bears them out.

Hobbes's lesson concerning what is necessary for salvation is a fairly simple one. He minimizes those requirements so as to minimize the chance for conflict between the heavenly sovereign and those of earth. "All that is NECESSARY to Salvatian [*sic*]," claims Hobbes, "is contained in two Vertues, *Faith in Christ,* and *Obedience to Laws.*"[34] In regard to the first of these virtues, faith,

Hobbes boils things down to one necessary article of faith: "The (*Unum Necessarium*) Onely Article of Faith, which the Scripture maketh simply Necessary to Salvation, is this, that JESUS IS THE CHRIST."[35]

When it comes to the laws individuals must obey to attain salvation, Hobbes continues his process of simplification and tries to separate from all of God's laws that pure minimum which is necessary for eternal salvation. Of the laws of nature, or of God, "the principall is, that we should not violate our Faith, that is, a commandment to obey our Civill Soveraigns, which wee constituted over us, by mutuall pact one with another."[36]

Hobbes's short, neat lesson on the question of how to obey both God and an earthly sovereign, therefore, teaches that only two things are necessary—faith that Christ is king, and obedience to civil sovereigns. But this lesson is not so neat after all, and it really has not resolved the fundamental tension between civil and divine authority, the source of martyrdom. Suppose the civil sovereign commands that one renounce one's faith in Christ—a faith that Hobbes claims is necessary for salvation—or else be put to death. Should one obey the civil sovereign, as the law of God commands, and renounce one's faith in Christ? Or should one follow the defiant advice Hobbes gave before starting this lesson, in which he claimed that it would be madness to obey a command which would result in eternal death?

Hobbes seems to realize that he had not cleared things up quite as much as he had intended, and at the very end of the chapter on the requirements for salvation, he discusses precisely that sort of situation described above. But Hobbes's advice at this point is no longer to disobey the civil sovereign. Hobbes now writes: "And for their *Faith*, it is internall, and invisible; They have the licence that Naaman had, and need not put themselves into danger for it."[37] One need not, therefore, become a martyr for one's faith, because faith is something internal. One no longer has to be willing to lay down one's earthly life, to consider that life worth nothing, in order to glorify God and his promise of eternal life.

I must point out that immediately after saying that one need not put oneself in danger for one's faith, Hobbes does explicitly recognize that martyrdom is still a possibility. People need not risk their earthly lives for their faith, writes Hobbes, "but if they do, they ought to expect their reward in Heaven, and not complain of their Lawfull Soveraign; much lesse make warre upon him. For he that is not glad of any just occasion of Martyrdome, has not the faith he professeth, but pretends it onely."[38] But Hobbes himself, I want to argue, is not glad for any just occasion of martyrdom. His theory of a Christian Commonwealth, based as it is upon his theory of the natural, human inclination to live in a commonwealth, does not leave much room for martyrdom or the traditional Christian justification of it.

To begin with, the context in which Hobbes offers his endorsement of martyrdom is one in which the objective seems to be preventing aggressive, dangerous martyrdom, such as that which occurs in religious wars. People need not become martyrs, but if they do, they should not "complain of" or "make war upon" their sovereign. In other words, if you must be a martyr, do it quietly, so that you do not disturb the peace of the commonwealth.

But while recognizing the possibility of nondisruptive martyrdom, Hobbes undermines the possibility of all forms of martyrdom by noting that the subjects whom he envisions in his commonwealth pose no real threat to sovereign authority and that there is no reason for any sovereign to persecute them. This is evident in the following quote, in which Hobbes tips his hand concerning the role of Christianity in the commonwealth.

> But what Infidel King is so unreasonable, as knowing he has a Subject, that waiteth for the second coming of Christ, after the present world shall bee burnt, and intendeth then to obey him, (which is the intent of beleeving that Jesus is the Christ,) and in the mean time thinketh himself bound to obey the laws of that Infidel King, (which all Christians are obliged in conscience to doe,) to put to death or persecute such a Subject?[39]

In a commonwealth formed of Hobbesian subjects, therefore, there is no need for martyrdom or persecution, even if the commonwealth is not a Christian one. The rational subject that Hobbes describes in the first part of *Leviathan* recognizes certain natural laws, but the general principle underlying all natural laws is that individuals cannot do that which threatens their earthly, mortal life. But martyrdom, as a voluntary relinquishment of one's time on earth, challenges this very idea of a law of nature, and is a threat to the rational order Hobbes tries to establish in *Leviathan*. Ultimately, Hobbes could not accept the possibility of martyrdom in modernity.

It should be emphasized that the view of human mortality that underlies the Hobbesian conceptions of the rational subject, the foundation and limitation of sovereign authority, and the danger of martyrdom, runs counter to the ambivalence I have identified in the Christian tradition. (See Chapters 2 and 6.) The Christian idea that earthly life, while being a gift from God, should nonetheless be held as something worthless and contemptible, something one would readily sacrifice, is not to be found in Hobbes. As I have tried to show here, Hobbes's theory of "a commonwealth ecclesiasticall and civil" is based on the preservation of one's life on earth, not one's eternal life in the "kingdom of God."

This attitude toward earthly life, which treats it as something inherently valuable and worth saving at any cost, characterizes modern asceticism and distinguishes it from its Christian predecessor. Given this perspective, it should be obvious why I treated Arendt's interpretation of Christianity so harshly in Chapter 2. While Arendt found a certain reverence for life in both Christianity and modernity, I identify a particular revaluation of mortal existence as a hallmark of modernity, which sets it apart from the Christian tradition. But this modern revaluation of earthly life is not the same attitude that Arendt located in modernity. It is not a reverence of the body and necessity, as Arendt's interpretation implies, but a transformation of the contemptuous dimension of Christianity's ambivalence toward the human condition.

The concern with prolonging or saving one's time on earth, which is so prevalent in Hobbes's *Leviathan*, is clearly opposed to the Christian tradition. But the other dimension of modern asceticism I have identified in Hobbes's thought—the desire for convenience or commodious living—also marks a transformation of that tradition, and not just because it increases, rather than restricts, consumption. The desire for a life of convenience, in which the limits of the body do not consume much of an individual's time, runs counter to the Christian idea that the toil and trouble of earthly life were a punishment for Adam's sin.

It should be noted here that the idea of saving time from necessity certainly did not emerge with modernity. As Arendt emphasized in her distinction between antiquity and modernity, the practice of slavery among the ancient Greeks freed adult males, to some extent, from the realm of necessity and provided them with time to participate in the *polis*. And Christianity, of course, developed its own practices in regard to the temporal requirements of necessity. But the regulation of time in Christianity—which began in the deserts of Egypt, was developed in the monasteries, and took a different shape in the worldly asceticism of the Puritans—differs from the attitude toward time and necessity which emerged in, or as, modernity.

The monastic and Puritannical regulations of time were, indeed, concerned with saving time from bodily necessity, but the objective of such regulation was to provide more time for prayers in the monastery or more time for the pursuit of a worldly calling for the Puritans. Those who saved and spent their time in the manner of Christian asceticism, in any of its variants, did so not in order to live a more comfortable, convenient life, but in order to attain eternal life. So it is not the need to save time from necessity alone which distinguishes modern asceticism from its Christian predecessors, but that need in conjunction with the modern attitude toward mortal, earthly life, which no longer treats that life as something to be endured as a punishment.

This distinction between modern and Christian asceticism, however, is not readily apparent in the thought of Hobbes. Hobbes does mention in *Leviathan* that "man was created in a condition Immortall, not subject to corruption, and consequently to nothing that tendeth to the dissolution of his nature; and fell from that happinesse by the sin of Adam."[40] But Hobbes makes this point in his discussion of salvation and eternal life in the third part of *Leviathan*. He does not use the idea of the fall to help explain the state of nature in the first part of the text, and therefore is not forced to come to grips with the conflict that exists between the rational idea that life in civil society can ameliorate the nasty, brutish aspects of earthly life and the Christian idea that those inconvenient aspects are the punishment for sin. Hobbes never tries to answer the question of how his idea of commodious living can be reconciled with the Christian idea of the fall. However, John Locke, another modern ascetic, struggles with this problem and, in his own manner, resolves it. An examination of Locke's treatment of the fall will help to make clear how this feature of modern asceticism—the desire or need for convenience—differs from the Christian tradition.

LOCKE AND CONVENIENCE

To begin with, I must point out that Locke's wrestling with Christian doctrine was not by any means limited to the idea of the fall. Like Hobbes, Locke sought to render Christianity a firm support for civil society, but Locke was more systematic than Hobbes in this effort. Years after publishing the famous *Two Treatises of Government* (1690), Locke offered his interpretation of the New Testament, or at least all that was reasonable in it, in *The Reasonableness of Christianity as Delivered in the Scriptures* (1695). And at the end of his life, Locke was working on paraphrases of Paul's epistles, in which he continued his rationalization of Christianity.[41] Locke's treatment of the fall, therefore, is just one facet of his project of constructing a reasonable Christianity, although it is an important facet.

Locke begins *The Reasonableness of Christianity* by squarely facing up to the problem that Hobbes never confronted in *Leviathan*. In the first sentence of this text, Locke acknowledges that there really is no way of avoiding the issue of the fall. "It is obvious to anyone who reads the New Testament," begins Locke, "that the doctrine of *redemption*, (and consequently of the gospel,) is founded upon the supposition of *Adam's fall*."[42] He continues: "To understand therefore what we are restored to by Jesus Christ, we must consider what the scripture shows we *lost by Adam*."[43]

In his consideration of what was lost by the fall, Locke offers an interpretation that initially appears to closely follow the traditional Christian interpretation. The punishment of Adam arose out of his violation of the command of God not to eat from a certain tree.[44] The punishment of which God warned Adam was death, but when Adam ate of the forbidden tree, "he did *not actually die*; but was turned out of paradise from the tree of life, and shut out for ever from it, 'lest he should take thereof and live for ever.'"[45] So the first thing Adam lost was immortality.

Locke's description of mortal life after the fall bears the temporal tone previously noted in Hobbes's description of life in the state of nature. On that very day that Adam ate, writes Locke, "His life began from thence to shorten and waste, and to have an end; and from thence, to his actual death, was but like the time of a prisoner between the sentence, and the execution which was in view and certain."[46] And Locke's theory of civil society, like Hobbes's, offers as a principal advantage to people the fact that their lives would be lengthened by consenting to give up some of their natural rights to form such a society.[47]

Along with his loss of eternal life, Adam also lost through his sin the bountiful life God had given him in paradise. Locke points out "that paradise was a place of *bliss*, as well as immortality; without toil, and without sorrow. But when man was turned out, he was exposed to the toil, anxiety, and frailties of this mortal life."[48] This is Locke's interpretation of God's cursing the ground in Genesis, and it is this aspect of the fall I want to stress in this discussion of Locke. For just as Locke's civil society shares with Hobbes's the feature that it can lengthen mortal life, it also offers the promise of alleviating some of the "toil, anxiety, and frailties of this mortal life." In his description of the state of nature, Hobbes calls these problems of mortal life "incommodities." Locke calls them "inconveniences."[49]

Now it is true that when Locke discusses the inconveniences of the state of nature, he, like Hobbes, does not cite the fall as the source of these problems, but follows Hobbes in citing instead the natural liberty and equality of people in that state.[50] But since each of these thinkers wants to use Christianity as a support for the commonwealth or civil society they are trying to build on natural, rational principles, there is occasion for asking how the convenient, commodious life they see as part of the promise of civil society can be reconciled with the Christian idea that mortal life is a punishment God inflicted upon humans.

In responding to this question, Locke begins to veer away from the traditional Christian interpretation of the fall. In Chapter 6, it will be recalled, I indicated that Augustine, Luther, and Calvin all recognized that the punishment inflicted upon Adam and Eve was extended to all their posterity.[51] Locke rejects this tenet of Christianity, and this rejection provides a way around this particular conflict between natural reason and Christian revelation.

On Locke's reading, Adam was indeed "*turned out* of paradise" as a punishment for his sin, but Adam's posterity was "*born out* of it."[52] It is not as a punishment that the generations after Adam were born out of paradise; it is rather just a condition in which they find themselves, given Adam's punishment. From Locke's perspective, to even suggest that God would have punished everyone for the transgression of Adam "when millions had never heard of, and no one had authorized to act for him—or be his representative"[53] violates not only the idea of consent, upon which Locke grounds his political philosophy, but it also is an affront to the honor of God. Such an idea, claims Locke, "indeed would be hard to reconcile with the notion we have of justice; and much more with the goodness and other attributes of the supreme being, (which he had declared of himself, and which reason and revelation must acknowledge to be in him.)"[54]

According to Locke, in order for God to punish Adam's posterity for his sin, God would have had to take away something to which that posterity had a right. That is the nature of punishment. But, Locke argues,

> The state of immortality in paradise [and, one could logically add, the blissful life there] is not *due* to the posterity of Adam, more than to any *other* creature. Nay, if God afforded them a temporary, mortal life, it is his gift; they owe it to his bounty; they could not claim it as their right; nor does he injure them when he takes it from them.[55]

Since people do not have a right even to their mortal life, much less can they claim to have a right to eternal, blissful life. Human mortality and the troubled life on earth, therefore, cannot be considered punishments, according to Locke. Even "such a temporary life as we now have, with all its frailties and ordinary miseries," writes Locke, "is better than *no being*."[56] It is just the condition in which individuals find themselves after Adam's fall, but it is still a gift from God.[57]

This interpretation of the fall has significant implications for that conflict between reason and revelation I have identified in the thought of Locke and Hobbes, but these implications are most readily apparent not in *The Reasonableness of Christianity*, but in an earlier work. When he attacked Filmer's *De Patriarcha* in the *First Treatise on Government*, Locke seems to already have understood the implications of his reading of the fall. Although Locke does not mention in *The Reasonableness of Christianity* the punishment God inflicted upon Eve, he does focus on that aspect of the fall in the *First Treatise* in order to refute one of Filmer's arguments for patriarchal hierarchy. And the interpretation of Eve's punishment that Locke offers in the *First Treatise* foretells of the reading of the fall Locke will offer later and also reveals the implications of this reading.

In the *First Treatise*, as in *The Reasonableness of Christianity*, Locke denies that God's punishment of Adam is intended for all generations.[58] However, Locke is not quite so restrictive when it comes to Eve's punishment. Locke writes about the words of Genesis 3:16, in which Eve's punishment of multiplied pain and subjection to her husband is inflicted, as follows: "And if we will take them as they were directed in particular to her, or in her, as a representative, to all other women, they will at most concern the female sex only, and import no more but that subjection that they should ordinarily be in to their husbands."[59] Locke's liberalism, obviously, was not as developed as that of J. S. Mill, since Locke suggests that all women should "ordinarily" be subject to their husbands on account of Eve. But Locke rejects the idea that even this subjection is a punishment which all women must suffer. Immediately following the preceding quote, Locke continues:

> But there is here no more law to oblige a woman to such a subjection, if the circumstances either of her condition or contract with her husband should exempt her from it, than there is that she should bring forth her children in sorrow and pain if there could be found a remedy for it.[60]

This stance toward the effect of Eve's punishment on all women reveals the route Locke takes to get around the constraints which the traditional interpretation of the fall posed to modern civil society. The punishments of Adam and Eve are not extended to all men and women as punishments, but rather simply as the condition in which they were born. As such, these inconveniences of earthly life need not be endured as some primal burden. Again referring to the words by which Eve was punished, Locke writes: "Neither will any one, I suppose, by these words think the weaker sex, as by a law so subjected to the curse contained in them, that it is their duty not to endeavor to avoid it."[61]

Rather than being bound to bear this condition, Locke argues that if a remedy could be found for the pain of childbirth or for the subjection of women to their husbands, women may certainly take steps to ameliorate these conditions or avoid them altogether. To extend Locke's argument to the condition in which humans in general find themselves after the fall, all the toil and trouble of earthly life can also be avoided. Men, no less than women, can avoid the brutishness and inconvenience of life in the state of fallen nature and, in fact, they form civil society precisely to avoid those conditions.

The rational or reasonable Christianity which Locke developed over the course of his career would appear to be better suited to the modern asceticism I have identified than it is to Christian asceticism itself. Locke's interpretation of the fall renders it compatible even with that very un-Christian facet of modern asceticism I stressed in my discussion of Hobbes—the idea that mortal, earthly life is something valuable and should be lengthened as much as possible. According to Locke's reading of the fall, people may be expected to avoid death just as they may be expected to avoid the other inconveniences of the state of nature. Mortality is no longer a punishment, but is rather merely a condition which may be avoided or changed.

In Locke's reasonable reading of Christianity, therefore, there no longer is any tension involved in the evaluation of earthly life. There also no longer appears to be any justification for Christian asceticism, which was based on the idea that inferior mortal life should willingly be sacrificed for immortal life. The Christian ascetic, aside from the priestly caste's attempts to gain power through difference, denied or regulated the needs and desires of the body so as to gain time for the contemplation of, and preparation for, an otherworldly immortality. With the revaluation of earthly, mortal life which was undertaken by Hobbes and Locke, the objective of Christian asceticism recedes into the background, and mortal, temporally finite life comes into prominence.

I believe it would be a mistake, however, to follow Arendt in reading modernity as an age in which life and the life processes are revered and held almost sacred.[62] On the contrary, my claim is that the modern attitude toward the body and necessity is still one of contempt, just as it was, in part at least, for Christianity. But this contemptuous dimension of modern asceticism tends to get shaded when one emphasizes how the Christian commonwealths offered by Hobbes and Locke differ from more traditional Christianity. Compared to the traditional Christian ambivalence toward earthly life, the Christianity espoused by Hobbes and Locke in their attempts to provide a religious foundation for the modern political order appears to be a celebration of mortal life and earthly comfort.

However, when one looks more closely at the direction in which modern consumption practices have developed, it becomes apparent that those practices continually assault the limits imposed by human embodiment. It is in this sense that the modern consumption of technology can be seen as a new way of denying the body, mortality, and necessity—as a new form of asceticism. Or to put this in Nietzsche's terms, the modern consumption of convenience is the latest form of homeopathic medicine, one administered by new priests—the scientists and technicians.

To get a better glimpse of modernity's contempt for the body, it is necessary to move beyond Hobbes and Locke, who still stand too close to Christianity to provide a clear image of modern asceticism. I will now turn to a very different line of thought, one developed in conscious contrast not only to liberal political thought, such as that of Hobbes and Locke, but also to Christianity. In the thought of Marx and many of those who have been influenced by that thought, there exists a more mature form of modern asceticism than that found in liberalism.

MARX AND NECESSITY

The ascetic dimension of Marx's thought is grounded in his attitude toward necessity. This attitude of Marx was most clearly stated very early in his career in what eventually came to be published as *The Economic and Philosophic Manuscripts of 1844* and has since become known as the "Paris Manuscripts." In these manuscripts, as in his later work, *Capital*, Marx challenges the liberal idea that through the exertion of one's labor one gains a property right to some of that which was given to all in common. (See note 12 of Chapter 4 for a presentation of the Lockean version of this conception of property.) In the earlier criticism of 1844, Marx introduces as a counter-concept to the liberal notion of property the idea of estranged, or alienated, labor,[63] and it is in his elaboration of this concept that Marx sets out his perspective on necessity.

According to Marx, each species has its own particular species character, which "is contained in the character of its life activity; and free, conscious activity is man's species character."[64] Under capitalism, however, people do not produce in a free and conscious manner, but work according to the dictates of capital, so they become alienated from the character of the human species. This is one of the ways in which individuals in a capitalist economy become alienated from their labor. As Marx puts it, "estranged labour estranges the *species* from man."[65]

But Marx goes further than simply noting this alienation of humans from their species character. Under capitalism, people actually come to resemble animals more than humans, because their productive activity takes on the species character of nonhuman animals. In distinguishing the species 'man' from animal species, Marx claims that:

> An animal only produces what it immediately needs for itself or its young. It produces one-sidedly, whilst man produces universally. It produces only under the dominion of immediate physical need, whilst man produces even when he is free from physical need and *only truly produces in freedom therefrom.*[66] (Emphasis added.)

Under the exploitative environment of capitalism, however, an individual's productive activity comes to be nothing more than "a means to his physical existence."[67] In other words, people work in order to satisfy the demands of their bodies, and in this they are more like animals than humans. Humans only "truly" produce, as humans, when they produce in the absence of physical needs.

In contrast to the life activity of animals, humans as a species have a more developed, richer life activity that satisfies needs higher and more refined than crude bodily needs. Marx believed that this higher nature of humans would eventually blossom once capitalism's exploitative, dehumanizing relations of production were demolished. "It will be seen," Marx wrote in the Paris Manuscripts, "how in place of the *wealth* and *poverty* of political economy come the *rich human being* and rich *human* need. The *rich* human being is simultaneously the human being *in need of* a totality of human life activities—the man in whom his own realization exists as an inner necessity, as *need.*"[68]

This attitude of Marx's toward necessity—in which the demands of the body are not just distinguished from richer human needs, but are denigrated to "mere" animal status—was not just a conviction Marx held early in his career before he undertook his painstaking analysis of the structure of capitalism. On the contrary, Marx expressed this same attitude in different contexts throughout his career. In *The German Ideology* (1845–46), which Marx wrote with Engels shortly after compiling the Paris Manuscripts, he again denigrates physical necessity, but this time not in distinguishing the human species from other animals, but in describing human productive activity after the proletarian appropriation of capitalism's highly developed forces of production. In the terms of the Paris Manuscripts, one might say that Marx and Engels are here describing the situation in which the species character of humans may be fulfilled: "Only at this stage does self-activity coincide with material life, which corresponds to the development of individuals into complete individuals and the casting-off of all natural limitations."[69]

By natural limitations, I read Marx and Engels to mean those conditions in which physical necessity, or the demands of the body, dominate human productive and consumptive activity. For just a page or so before making the preceding claim about casting off natural limitations, Marx and Engels note:

> The only connection which still links [the majority of individuals] with the productive forces and with their own existence—labour—has lost all semblance of self-activity and only sustains their life by stunting it.... finally material life appears as the end, and what produces this material life, labour, (which is

> now the only possible but, as we see, negative form of self-activity), as the means.[70]

In other words, people have become like animals; their life activity is nothing more than the means of existence.

Marx eventually came to recognize that capitalism did not simply reduce human needs to the level of animals. Even though capitalism undoubtedly had this dehumanizing effect, by 1857–58, when he wrote those notebooks which eventually were published as *Grundrisse*, Marx recognized that capitalism also had a tendency to develop and enrich human needs. In *Grundrisse*, Marx claimed that: "Capital's ceaseless striving toward the general form of wealth drives labour beyond the limits of its natural paltriness...and thus creates the material elements for the development of the rich individuality which is as all-sided in its production as in its consumption."[71]

But still, Marx set this "rich individuality" against the demands of the body. He continues his description of the richly producing and consuming individuality by saying that its "labour also therefore appears no longer as labour, but as the full development of activity itself, in which natural necessity in its direct form has disappeared."[72] Here, as in the Paris Manuscripts and *The German Ideology*, the demands of the body must be overcome as limitations on human activity if humans are to attain the complete individuality of their species character.

Marx's attitude toward necessity, which denigrates the demands of the body and celebrates the creation of rich human needs, appears to be a reversal of that needs/wants distinction identified earlier in the thought of Luther and Franklin, or more accurately, Poor Richard. Unlike those earlier perspectives on necessity, which sought to limit the multiplication of needs by granting to the 'real,' physical needs a certain predominance, Marx held that the creation of new needs, some of which I would describe as limits of the body, depended on humanity's moving beyond the realm of "crude practical need."[73] For Marx, 'real' needs were not a limit to be imposed upon the development of earthly conveniences; they were, instead, an obstacle to such development which must be overcome.

It must also be emphasized that Marx deeply appreciated

and applauded the expansive effect that capitalism had on necessity. He shared with the liberals Hobbes and Locke an appreciation for the way in which a civil society based on private property (i.e., a capitalist society) helped to develop, advance, or civilize people. In the following long quote from *Grundrisse*, the laudatory tone of Marx's description of this particular feature of capitalist production reveals this appreciation, as well as the threat such production poses. All of the following, claims Marx, "is a condition of production founded on capital":

> Hence exploration of all nature in order to discover new, useful qualities in things; universal exchange of the products of all alien climates and lands; new (artificial) preparation of natural objects, by which they are given new use values. The exploration of the earth in all directions, to discover new things of use as well as new useful qualities of the old; such as new qualities of them as raw materials etc.; the development, hence, of the natural sciences to their highest point; likewise the discovery, creation and satisfaction of new needs arising from society itself; the cultivation of all the qualities of the social human being, production of the same in a form as rich as possible in needs, because rich in qualities and relations—production for this being as the most total and universal possible social product, for, in order to take gratification in a many-sided way, he must be capable of many pleasures...hence cultured to a high degree...[74]

The attitude Marx displays in the preceding passage indicates how the unlimited development of needs in modernity helps to foster the technological appropriation of the world. Marx's juxtaposition of ideas such as "the discovery, creation, and satisfaction of new needs" and "the exploration of all nature in order to discover new, useful qualities in things," reveals how Marx's "social human being...as rich as possible in needs," participates in the modern project of Enframing. (See Chapter 1, note 1, and Chapter 4, note 37.) But once again, my argument is not concerned with the expansive nature of technology, but rather with the attitude toward the body and the self that underlies the consumption of technology in modernity. Marx's endorsement of the proliferation of needs or the expansion of necessity, I want to suggest, shares with Hobbes's and Locke's conceptions of civil society that modern ascetic tendency to deny the body, or the temporal-spatial limits which are imposed by it.

In fairness to Marx, I must mention that later in his career, when he was working on the third volume of *Capital*, he appears to have moved away from his earlier position on necessity to some degree and was willing to accept limitations on the development of needs. But the limitations Marx had in mind were still centered on minimizing the time that had to be spent on bodily necessity.

In this late discussion of necessity, Marx maintains his earlier position, that "the realm of freedom really begins only where labour determined by necessity and external expediency ends."[75] However, Marx recognizes at this point that physical necessity is not a static concept, but that even such basic needs are susceptible to development. As Marx puts it, "the realm of natural necessity expands with his [man's] development, because his needs do too."[76] The limitations of needs Marx had in mind were centered on this realm of natural necessity, as becomes apparent in the following quote. After noting that the realm of natural needs expands, Marx continues:

> but the productive forces to satisfy these expand at the same time. Freedom, in this sphere, can consist only in this, that socialized man, the associated producers, govern the human metabolism with nature in a rational way, bringing it under their collective control instead of being dominated by it as a blind power; accomplishing it with the least expenditure of energy and in conditions most worthy and appropriate for their human nature.[77]

As truly human producers, therefore, people must collectively and rationally control the development of needs. This recognition of the need for some limitation on needs is an important development in Marx's thought. But as I read this last passage, Marx still appears to be claiming that freedom requires that even developed needs, at least those which are related to physical, bodily necessity, must be satisfied in the fastest, easiest, most convenient manner. In other words, the "conditions most worthy and appropriate for...human nature" are the conditions in which the limits which are imposed by the body interfere with the use of time in a minimal manner. The irony of Marx's necessity/freedom distinction, however, is that the value of convenience, which underlies Marx's conception of necessity from its earliest form in the Paris Manuscripts to its last formulation in *Capital*, thwarts the promise of freedom which Marx offers.

At the conclusion of his mature discussion of necessity, Marx offers a brief glimpse of the freedom he has in mind for a truly human productive environment. Of the realm of necessity, Marx writes: "The true realm of freedom, the development of human power as an end in itself, begins beyond it, though it can only flourish with this realm of necessity as its basis. The reduction of the working day is the basic prerequisite."[78]

Marx anticipated that increases in productivity would eventually provide this prerequisite for freedom by reducing the amount of time people had to spend satisfying their needs through work. But this expectation no longer appears to be well-founded in either socialist or capitalist economies. Of course, there is no doubt that the length of the working day has indeed been shortened from the brutal hours Marx and Engels witnessed in the British textile mills of the nineteenth century, and the eight-hour day has become the standard in the most advanced economies. But in the case of the United States, this standard has been in place since 1938 and has not been reduced through half a century of improvements and refinements in the forces of production.

As I read him, Marx meant by the reduction of the working day a continuous process which would free up more and more time from the realm of necessity as the forces of production were developed. (Marx did emphasize, however, that necessity would never be eliminated altogether.[79]) In my judgment, this reduction in the length of the working day has not come to pass in either the East or the West, and it does not look as though it is about to. There has not yet emerged in modernity a way of being that has provided a realm of freedom in which individuals could fully develop their human faculties. In fact, it seems to me that moderns in both the East and the West are ready to work harder and longer in order to satisfy the continually expanding list of needs—or more accurately, to satisfy the need to continually expand that list.

Explanations of Marx's unfulfilled expectations concerning the length of the working day are certainly available. In the West, the control that capital exerts on the development of needs, and in the East, the corruption and inefficiency of the bureaucratic state, are obvious choices of explanations which could be developed at length by those so inclined. What I would like to suggest as an explanation of this shortcoming of Marx's thought, however, is that humans have not come to find any higher, richer needs which exist beyond the expanded realm of physical need. Or to put this differently, the higher value upon which modernity has settled is the value of convenience, and this value is satisfied through the continued labor of people in the capitalist or socialist production process and the increased consumption of technological apparatuses.

In an ironic sense, Marx's description of the "true realm of freedom" has contributed to the failure of modernity to provide the very conditions necessary for that realm. The need to continually reduce the amount of time and energy people expend upon bodily necessity has become the end in itself. And as I tried to show earlier, such assaults on the limits of the body, especially temporal limits, are neverending. It is no wonder that modern ascetics have little time for "the development of human powers as an end in itself." They are too busy trying to be rid of their bodies and the limits those bodies impose on their freedom. Marx's conceptions of freedom and necessity only contribute to, and cannot resolve, this dilemma.

Marx's attitude toward necessity still exerts an influence on contemporary Marxism. A very similar attitude toward necessity can be found among some of the texts I examined in Chapter 3. Preteceille and Terrail's book on capitalism, consumption, and needs, and Mandel's analysis of late capitalism both bear the influence of the Marxian concept of necessity, and they do so in a similar manner. Both texts employ the Marxist distinction between base, basic needs and higher, richer ones, and they use this distinction in order to forestall any discussion about limiting the development of needs.

I mentioned earlier that Preteceille and Terrail call for "the expansion, development and transformation" of the existing needs of workers, and that they claim that "breaking up capitalist hegemony entails an explosion of needs."[80] Now the context in which Preteceille and Terrail make this claim indicates that the new needs they have in mind are not limits of the body, but are instead needs having to do with political and economic power. But in other places in their essays, Preteceille and Terrail do indicate that they consider modern developments in physical necessity to be beyond reproach. In one essay, Preteceille claims that the

> primacy of living labour cannot be achieved, as it is sometimes suggested today, by denouncing science and technology as the guilty parties.... Nor can it be achieved by a return to earlier forms of production, to more 'human' forms of craft and pastoral organization.... On the contrary, it is scientific and technological progress that can help to provide the answer.[81]

But it's not just that science and technology can be of help in providing the answer. From Preteceille and Terrail's perspective, science and technology really are not part of the problem. Preteceille goes so far as to say that to question science and technology and all the productive and consumptive development they have provided, to argue for a "regression of productive forces, is nothing but an individualist flight on the part of the petit-bourgeois elements edged out by the crisis."[82] And in another essay, Preteceille describes as "regressive and illusory" any attempts to solve "problems of energy, transport or ecology, [by] suggesting a return to archaic forms of production, and denouncing technology in general rather than its form and use under capitalism."[83]

This same attitude toward modern necessity is present in Mandel's *Late Capitalism.* After listing various sources of "consumer society," Mandel claims:

> Any rejection of the so-called 'consumer society' which moves beyond justified condemnation of the commercialization and dehumanization of consumption by capitalism to attack the historical extension of needs and consumption in general (i.e., moves from social criticism to a critique of civilization), turns back the clock from scientific to utopian socialism and from historical materialism to idealism.[84]

To be fair, it must be noted that Mandel does recognize that "the possibilities of developing and differentiating material consumption cannot be unlimited," but he does not get into that in *Late Capitalism*.[85] But if Mandel is willing to accept limits on the development of needs, these limits have not yet been reached. With consumer society, claims Mandel, there has occurred a "genuine extension of needs,"[86] and Mandel, I believe, anxiously anticipates the continuation of this 'progress.' The needs themselves are not open to question; the problem with consumer society lies solely in the relations of production, not in the values which guide and direct that consumption.

Mandel explicitly cites Marx as the source of this attitude toward consumer society or, as I would put it, modern necessity. He continues the preceding quote by saying: "Marx fully appreciated and stressed the civilizing function of capital, which he saw as the necessary preparation of the material basis for a 'rich individuality.'"[87] Very well, but Marx wrote over a hundred years ago. Mandel, as well as Preteceille and Terrail, cannot fathom the query of whether modern individuals might not be over-prepared for that rich individuality. Such a question would most likely appear to them "regressive and illusory," or perhaps even "vulgar and mean."[88] Nonetheless, this is one way of formulating the question this text attempts to pose.

This chapter on the traces of modern asceticism that can be found in liberal and radical political thought began with a discussion of the role which death and mortality play in Hobbes's *Leviathan*. The evasion of death upon which Hobbes's civil society is based was presented as one element of modern asceticism. Since that point, the discussion shifted to the other element of modern asceticism, the saving of time from necessity. In the final chapter of this text I will return to the issue of death, because it is around this troubling issue that the technological trajectory of modernity crystallizes and reveals itself most clearly as a continuation of the ascetic denial of the body.

CHAPTER 9

The End of Death

The issue of human mortality has assumed a position of primary importance in the twentieth century. The mortality of the species has been brought to our attention first by the threat of nuclear annihilation, then nuclear refrigeration, and recently environmental trends have been noted that indicate the threat may come instead from the gradual warming of the planet. But prior to these widely announced calls to reflection on the possibility of the death of the species, there was another very different call to reflect upon the possibility of death. This call was issued by Martin Heidegger in *Being and Time* (1927),[1] where he presents the possibility of death not as a danger to be avoided or averted, but as something to be heeded. However, the call to reflect upon mortality and the unique attitude toward death that Heidegger offers in *Being and Time* have gone unheeded. But in these reflections on death, a way is revealed for avoiding those dangers posed by the technological order of modernity. In Heidegger's stance toward mortality there lies a challenge to modern asceticism.

Heidegger spends the first half of *Being and Time* undertaking what he calls a "Preparatory Fundamental Analysis of Dasein." (*Dasein* is literally translated as 'being there,' and is the term Heidegger uses to reveal the being of persons.[2]) The outcome of this preparatory analysis of *Dasein* is the identification of 'care' as the being of humans, but this analysis is not completely satisfactory to Heidegger because it has failed to reveal the whole of *Dasein.* Even though Heidegger suspects that the being of humans is precisely the sort of being that cannot be grasped in its totality, he nevertheless asks whether "we" have "indeed exhausted all the possibilities for making Dasein accessible in its wholeness?"[3]

In the second half of the text, Heidegger follows another path toward grasping the totality of human being, one concerned with the essential temporality of that being, and it is here that he raises the issue of death. For Heidegger, the fact that human being is mortal, that it has an end, is what renders *Dasein* an inherently temporal being. *Dasein* is the sort of being that always exists in the manner of looking ahead toward future possibilities. As Heidegger puts it, "the primary item in care is the 'ahead-of-itself'."[4] And the one undeniable possibility which lies before *Dasein* is that of death. Death is the 'not yet' that always lies ahead of *Dasein* and renders *Dasein* a temporally concerned being. So, in order to more fully understand human being, Heidegger claims that "we have the task of characterizing ontologically *Dasein*'s Being-at-an-end and of achieving an existential conception of death."[5]

Toward this end, Heidegger first examines the possibility of one person experiencing the death of another, but finds that such an experience cannot provide an authentic existential understanding of death. "The dying of Others is not something which we experience in a genuine sense; at most we are always just 'there alongside'."[6] A genuine experience of death can only be attained by experiencing death as one's own. "Death is a possibility-of-Being which *Dasein* itself has to take over in every case," claims Heidegger. "With death, *Dasein* stands before itself in its ownmost potentiality-for-Being."[7]

What Heidegger has in mind as an authentic experience of death is not the occasional recognition one may have that one is mortal and that at some future point one's life will end. Death, as a possibility-of-Being, is not a distant event. "On the contrary," writes Heidegger, "if *Dasein* exists, it has already been *thrown* into this possibility.... Factically, *Dasein* is dying as long as it exists."[8] This understanding of death as one's ownmost potentiality can only be attained when one experiences what Heidegger calls "anxiety" in the face of death. But he cautions that "anxiety in the face of death must not be confused with fear in the face of one's demise. This anxiety is not an accidental or random mood of 'weakness' in some individual; but, as a basic state-of-mind of *Dasein*, it amounts to the disclosedness of the fact that *Dasein* exists as thrown Being *towards* its end."[9]

Heidegger recognizes that, in ordinary circumstances, people remain oblivious to the ineluctable possibility of death and that "evasive concealment in the face of death dominates everydayness."[10] With a certain amount of contempt, Heidegger claims that "the 'they' does not permit us the courage for anxiety in the face of death."[11] Nevertheless, Heidegger does outline what he describes as an "authentic Being-towards-death," in which anxiety is fully experienced, and the individual is "wrenched" away from the everyday world of the "they."[12]

It should be noted that this authenticating function of death is quite different from the role death plays in Hobbes's thought, even though death is as central to Hobbes's scheme in *Leviathan* as it is to Heidegger's *Being and Time*. For Hobbes, it was the threat of violent, premature death that impelled men out of the state of nature and into civil society. In the everyday circumstances of civil society the immediate threat of death, that most "incommodious" feature of the state of nature, was minimized as death was pushed off into the distance, and the homeopathic medicine of convenience was then administered to numb these subjects to their distant mortality. In *Being and Time*, on the other hand, Heidegger strives to dispel the "constant tranquilization about death"[13] that occurs in the everyday world, and to bring death back into focus.

When I say that Heidegger wants to bring death back into focus, I do not mean in the manner of Christians such as Calvin, who urged his followers to "ardently long for death, and constantly meditate upon it." Heidegger explicitly states that authentic Being-towards-death "cannot have the character of concernfully Being out to get itself actualized..." and "neither can we mean 'dwelling upon the end in its possibility.'"[14] The experience Heidegger has in mind is not a longing for one's demise, nor is it a pondering of the possible circumstances of one's death. Rather, the authentic understanding of death "penetrate[s] into it *as the possibility of the impossibility of any existence at all*...of the impossibility of every way of comporting oneself towards anything, of every way of existing."[15] This existential impossibility of any existence, of course, precludes the possibility of any existence in an otherworldly afterlife.[16]

What is provided to *Dasein* by this authentic experience of death is not a fear that causes one to tremble in the very core of one's being, nor is it the promise of one's eventual release from the burdens of this world. What authentic Being-towards-death provides is freedom. Through this experience, "one is liberated from one's lostness in those possibilities which may accidentally thrust themselves upon one; and one is liberated in such a way that for the first time one can authentically understand and choose among the factual possibilities lying ahead of that possibility which is not to be outstripped"[17]—i.e., death. Heidegger is emphatic about this liberating potential of death; he summarizes his characterization of authentic Being-toward-death in bold, block letters as an "impassioned **freedom for death.**"[18]

This stance toward death which Heidegger tries to articulate is fundamentally opposed to the attitude toward death found in modern asceticism. For Hobbes, death was the ultimate temporal limit and, as such, it had to be avoided at any cost. For Heidegger, on the other hand, death is a limit toward which one should strive to remain open. "In anticipating the indefinite certainty of death," writes Heidegger, "Dasein opens itself to a constant *threat* arising out of its own 'there'. In this very threat Being-towards-the-end must maintain itself."[19] In this acceptance of and openness to the fundamental limit of mortality, Heidegger's thought provides an antidote to the "narcotics" of the ascetic priests. For Heidegger, death is not simply a bodily limit that must be overcome.

Regrettably, Heidegger's challenge to modern asceticism has not been taken up by many contemporary philosophers, much less by modern culture. Rather, the fear and evasion of death which Hobbes placed at the heart of modernity has flourished in the twentieth century and has become so highly developed that a new conception of immortality has emerged. This development of modern asceticism reveals itself clearly in the writings of a thinker who was influenced not only by Heidegger but by Marx and Freud as well. Herbert Marcuse's stance toward mortality and necessity exemplifies this late phase in the development of modern asceticism.

Before examining Marcuse's writings on death, I must point out that, although Marcuse was influenced to some degree by Marx and accepted Marx's freedom/necessity dichotomy,[20] he nevertheless took an important step beyond the Marxist conception of necessity. In regard to the dynamic development of needs in modernity, Marcuse was not as sanguine as were those Marxists examined at the end of the previous chapter. In *An Essay on Liberation*, Marcuse explicitly acknowledges that the proliferation of needs under modern capitalism helps to defuse the revolutionary potential of those who are alienated and exploited by capital. As Marcuse puts it:

> What is now at stake are the needs themselves. At this stage, the question is no longer: how can the individual satisfy his own needs without hurting others, but rather: how can he satisfy his needs without hurting himself, without reproducing, through his aspirations and satisfactions, his dependence on an exploitative apparatus which, in satisfying his needs, perpetuates his servitude?[21]

Marcuse calls for "the ascent of needs and satisfactions very different from and even antagonistic to those prevalent in the exploitative societies"[22] and anticipates that such a qualitative shift in necessity will usher in an expansion of freedom. In *Eros and Civilization*, Marcuse claims that this "expanding realm of freedom becomes truly a realm of play—of the free play of individual faculties. Thus liberated, they will generate new forms of realization and of discovering the world, which in turn will reshape the realm of necessity, the struggle for existence."[23] Marcuse foresaw the development of a "new sensibility," of an aesthetic appreciation or appropriation of the world that would challenge the instrumental approach of modern industrial civilization.[24] To this extent, Marcuse has moved beyond the simple celebration of technological development that can be found in many other writers who were influenced by Marx.

But even though Marcuse's attitude toward needs may mark a break with the ascetic compulsion to develop new techniques for saving time from necessity, Marcuse carried the other dimension of modern asceticism—the evasion of death—to a more refined stage. Marcuse articulates his attitude toward death in *Eros and Civilization* and in an essay entitled "The Ideology of Death," both of which were written in the 1950's. In the former, Marcuse, like Heidegger, describes death in temporal terms as "the final negativity of time."[25] But as such, death for Marcuse is the ultimate limit to human freedom; it is the "one innermost obstacle [which] seems to defy all project of a non-repressive development."[26] The reason Marcuse sees death as such a spoiler to the goal of human freedom is that "the mere anticipation of the inevitable end, present in every instant, introduces a repressive element into all libidinal relations and renders pleasure itself painful."[27] The difference between this attitude toward death and Heidegger's, which finds freedom precisely in the experience of anxiety in the face of death, should be obvious.

Now even though Marcuse moves beyond much contemporary Marxism in his conception of a transformed necessity, his idea that death is the ultimate obstacle to nonrepressive development is still quite similar to Marx's attitude that necessity should be overcome and minimized through technological developments. This similarity becomes unmistakable in "The Ideology of Death," where Marcuse discusses necessity as follows: "Necessity indicates lack of power: inability to change what is—the term is meaningful only as the coterminus of freedom."[28] And death, as the ultimate bodily necessity, is a limit to freedom. But Marcuse emphasizes that death does not enjoy any special status in the realm of necessity simply because it is the final limit. As he puts it, death is simply "a technical limit of human freedom."[29] It is nothing more than one facet of bodily necessity, just another limit the body imposes on human freedom.

Marcuse's claim is that overcoming or surpassing the limit of death "would become the recognized goal of the individual and social endeavor,"[30] if only the prevailing attitude toward death, "the ideology of death," could be overcome. By the ideology of death, Marcuse is referring to the use to which he claims death has been put throughout the history of Western philosophy, beginning with Socrates and culminating in Heidegger. This ideology renders the issue of death a support for existing political orders by inverting death, as a bit of biological necessity, into death as the end, or *telos*, of human life. Through this "ontological inversion," Marcuse claims, "a brute biological fact, permeated with pain, horror and despair, is transformed into an existential privilege."[31] Death comes to be treated as pertaining "to the essence of human life, to its existential fulfillment;" death becomes "the very token of...freedom."[32]

What bothers Marcuse, however, is not merely the fact that in the ideology of death something brutish and basic has become ontologically and existentially significant. Along with this, the acceptance of the necessity of death plays into the hands of the dominant forces in society. In *Eros and Civilization*, Marcuse writes:

> Whether death is feared as constant threat, or glorified as supreme sacrifice, or accepted as fate, the education for consent to death introduces an element of surrender into life from the beginning—surrender and submission. It stifles 'utopian' efforts. The powers that be have a deep affinity to death; death is a token of unfreedom, of defeat.[33]

And in "The Ideology of Death," Marcuse claims that "no domination is complete without the threat of death and the recognized right to dispense death...no domination is complete unless death, thus institutionalized, is recognized as more than natural necessity and brute fact, namely as *justified* and *justification*."[34]

As a counter to this submissive ideology of death, Marcuse offers "some kind of 'normal' attitude toward death—normal in terms of the plain observable facts."[35] What Marcuse has in mind here, of course, is that attitude toward death which sees it as a limit of the body to be overcome through technological development. According to this normal attitude, death will become "a necessity against which the unrepressed energy of mankind will protest, against which it will wage its greatest struggle."[36]

From the perspective offered in my text, Marcuse's choice of the word "normal" to describe that non-ideological attitude toward death is a fortuitous one, and it is a choice with which I fully agree. As I indicated earlier, that same attitude was present in both Hobbes and Locke. Recall that both of these theorists found as one of the primary benefits of civil society the fact that it could prolong life by allowing its members to avoid the risk of premature death, a risk which prevailed in the state of nature. Marcuse's attitude toward death as a technical limit, therefore, is normal not just in the sense that it is based on "plain observable facts" (whatever they might be), but is also normal in the sense that it conforms to the ascetic norms of modernity. Marcuse's goal of overcoming the brutish, biological necessity of death can be interpreted as the culmination of modernity's attempt to deny the body and its limits.

In regard to the limit of death, modernity has become adept at not only prolonging life, through the development of technologies such as organ transplants and life-support systems. The limit of death has also been pushed back at the other end, as the point of "fetal viability" has receded in the face of technological developments. And if I may speculate on what the future holds for humanity's struggle against death, I think that death will eventually be overcome as a limit, and the body will indeed be eliminated as the source of limits or necessity. However, these so-called "achievements" will only be won at the cost of a greater dependency on the technological order, socialist or capitalist, which provides them. Contrary to Marx's and Marcuse's expectations, these victories over the body have not engendered "new forms of realization and of discovering the world;" they have not fostered a "new sensibility" or an "art of living." (See note 24 of this chapter.)

At this point another perspective on the role that death plays in modernity may be instructive. Michel Foucault offers a fundamentally different interpretation from Marcuse's of the role death plays in maintaining relations of domination. According to Foucault, the right to inflict death is no longer the primary source of relations of power, as Marcuse argues. Foucault admits that, in classical relations of power, "the sovereign exercised his right of life only by exercising his right to kill, or by refraining from killing."[37] But a "very profound transformation of these mechanisms of power" has occured since the classical age, Foucault argues, and power is now exercised primarily through the administration of life, rather than the imposition of death. As Foucault puts it, "this formidable power of death...now presents itself as the counterpart of a power that exerts a positive influence on life, that endeavors to administer, optimize, and multiply it, subjecting it to precise controls."[38]

Viewed from this Foucauldian perspective, the immortality that Marcuse desires and modernity is on the verge of providing no longer appears as the culmination of freedom. Instead, the technological victory over death may indeed open up a completely new realm of order in which humans, or at least part of them, will be subjected to even greater control and regulation. Here I will enlist the support of a writer much more attuned to technological development than I am and use his celebratory vision of the future to help make these suspicions clearer. The following is from Robert Jastrow's *The Enchanted Loom: Mind in the Universe*:

> At last the human brain, ensconced in a computer, has been liberated from the weaknesses of mortal flesh. Connected to cameras, instruments and engine controls, the brain sees, feels, and responds to stimuli. It is in control of its own destiny. The machine is its body; it is the machine's mind. The union of mind and machine has created a new form of existence, as well designed for life in the future as man is designed for life on the African savanna.
>
> It seems to me that this must be the mature form of intelligent life in the Universe. Housed in indestructible lattices of silicon, and no longer constrained in the span of its years by the life and death cycle of a biological organism, such a kind of life could live forever. It would be the kind of life that could leave its parent planet to roam the space between the stars. Man as we know him will never make that trip, for the passage takes a million years. But the artificial brain, sealed within the protective hull of a star ship, and nourished by electricity collected from starlight, could last a million years or more. For a brain living in a computer, the voyage to another star would present no problems.[39]

In this vision of the future, the various strains of modern asceticism are carried to new heights. The temporal limits of the body are shattered as the brain abandons this mortal "biological organism," and no longer has to look ahead toward the 'not yet' of death. The time that had been spent satisfying the needs of the body is also eliminated as the brain is "liberated from the weaknesses of mortal flesh;" the disembodied mind is nourished, without time or effort, by starlight. Along with these temporal limits, the spatial constraints which the body imposed are also overcome, and the brain is finally free to travel among the stars.

But it is not clear that these developments will mark the end of necessity or even the development of qualitatively new needs. Why settle for a million-year life-span? That limit can surely be pushed back. And as for the time it takes to travel between stars, it will certainly be necessary to continually shorten this. There really will be no end to the development of needs in space, just as there have been none here on earth. Humans, or at least the brains of these beings, will continually struggle against the imposition of any ends or limits. To that extent, even these disembodied, extraterrestrial beings will resemble their mortal predecessors, who were unable to accept any indication of human finitude.

But more importantly, there is certainly room to question whether the disembodied star traveler will experience any greater freedom than its mortal predecessor. While this interstellar traveler may be free from the body, it will certainly be dependent on the technological order that provides the means to leave that body behind. On the American frontier there were possibilities available for resisting the technological order which was becoming established; in outer space, however, there would appear to be little room for squatters who refuse to abide by the dictates of the technological order which makes such an existence possible.

For those who do refuse or are unable to participate in that order and remain bound to the earth and their bodies, it is not likely that they will be eliminated as a threat through outright violence and repression. Rather, the technological order will, to use Foucault's terms, "disallow" this new form of life to such people "to the point of death."[40] While the technological order establishes the means to allow these new beings to escape the limitations of the body and earth, the various technologically induced calamities which threaten the human species will quite likely play themselves out, making embodied life on the planet impossible. It is in this sense that life will be disallowed to the point of death.

Any attempt to resist this trajectory of modernity, which is heading toward life among the stars, can benefit from the lesson Heidegger offered early in this century. The careful, concerned stance toward death Heidegger presents in *Being and Time* could be expanded beyond that ultimate end and cultivated as an attitude toward all manifestations of human finitude. The time spent satisfying the demands of the body as well as the earthly environment in which humans dwell could become objects of attention and care, rather than technical limits which must be shattered. Such an acceptance of limits could help to challenge the threat which technical culture poses to the earth and to bodily existence.

Of course, such an acceptance of limits can certainly be portrayed as a form of asceticism in its own right. Abandoning the pursuit of convenience may indeed appear as a form of self-denial, especially to techno-fetishists. I have no qualms with such a characterization of this attitude toward bodily necessity, as long as the distinction between this and modern asceticism is noted. Just as there was an important difference between the *askesis* of the Greeks[41] and the asceticism of those early Christians who left the fertile delta of the Nile and moved beyond the boundary of tombs into the inhuman environment of the desert,[42] there is a crucial difference between these more recent forms of asceticism. If the attitude toward necessity I am endorsing can indeed be described as asceticism, it is an asceticism which breaks with the contempt for the body that underlies both Christian and modern asceticism. This asceticism promotes an appreciation for the body and its limits, while the modern asceticism I have criticized seeks to deny the spatio-temporal limits of the body and escape into the desert of space.

NOTES

CHAPTER 1. INTRODUCTION

1. See Jacques Ellul, *The Technological Society*, trans. John Wilkinson (New York: Vintage Books, 1964), and Martin Heidegger, *The Question Concerning Technology and Other Essays*, trans. William Lovitt (New York: Harper & Row, 1977), especially "The Question Concerning Technology" and "The Age of the World Picture."

In *The Technological Society*, Ellul traces the dissemination throughout society of "technique." Of technique, Ellul writes that it "constructs the kind of world the machine needs and introduces order where the incoherent banging of machinery heaped up ruins. It clarifies, arranges, and rationalizes; it does in the domain of the abstract what the machine did in the domain of labor. It is efficient and brings efficiency to everything" (p. 5).

In the first of Heidegger's essays previously cited, Heidegger characterizes modernity as an age in which "everywhere everything is ordered to stand by, to be immediately at hand, indeed to stand there just so that it may be on call for a further ordering" (p. 17). Heidegger emphasizes that this ordering is not merely "a human handiwork," but that man too is "challenged forth" by this ordering (pp. 18-21).

2. Michel Foucault, for one, appreciated the dimensions of the Iranian rejection. In an interview published under the title "Iran: The Spirit of a World Without Spirit," Foucault said of the Iranian Revolution that it was a "radical rejection: the rejection by a people, not only of foreigners, but of everything that had constituted, for years, for centuries, its political destiny." Michel Foucault, *Politics, Philosophy, Culture: Interviews and Other Writings, 1977–84*, ed. Lawrence D. Kritzman, trans. Alan Sheridan et al. (New York: Routledge, Chapman & Hall, 1988), p. 215.

But Foucault emphasizes that the Iranians sought to transform not only their government; they sought to transform their very being as well.

> In rising up, the Iranians said to themselves—and this perhaps is the soul of the uprising: 'Of course, we have to change this regime and get rid of this man, we have to change this corrupt administration, we have to change the whole country, the political organization, the economic system, the foreign policy. But, above all, we have to change ourselves. Our way of being, our relationship with others, with things, with eternity, with God, etc., must be completely changed and there will only be a true revolution if this radical change in our experience takes place.' I believe that it is here that Islam played a role.... there was the desire to renew their entire existence by going back to a spiritual experience that they could find within Shi'ite Islam itself. (pp. 217–18)

3. Examples of social resistance are found in the preceding paragraph. An example of what I have in mind when I refer to resistance in nature is the apparent inability of modern physics to set in order the activity of flowing liquids. The study of fluid mechanics has so far been unable to establish any order in the movement of fluids when they travel over a variable surface, such as a streambed. This movement is simply chaos from the perspective of fluid mechanics. This is not to say, of course, that physicists are not doggedly trying to uncover the regularity, and hence predictability and controllability of the chaotic phenomenon of nature. See James Gleick, *Chaos: Making a New Science* (New York: Penguin Books, 1987).

4. Max Scheler, *Problems of a Sociology of Knowledge*, trans. Manfred S. Frings (London: Routledge & Kegan Paul, 1980), pp. 100–139. Scheler summarizes this point in a footnote: "To think of the world as value-free is a task that man posits to himself for the sake of value: the vital values of *mastery* and power over things" (p. 127, note 86).

5. Friedrich Nietzsche, *The Will to Power*, trans. Walter Kaufmann and R. J. Hollingdale (New York: Vintage Books, 1967), Sec. 588, p. 322.

6. Scheler, *Problems of a Sociology of Knowledge*, p. 102. Compare Scheler's statement on the primacy of values and valuation with the following passages from Nietzsche, *The Will to Power*: "In valuations are expressed conditions of preservation and growth. All our organs of knowledge and our senses are developed only with regard to conditions of preservation and growth.... 'The *real* and *apparent* world'—I have traced this antithesis back to *value* relations" (Sec. 507, pp. 275–6); and "It cannot be doubted that *all sense perceptions are permeated with value judgments*" (Sec. 505, p. 275).

7. John Raphael Staude, *Max Scheler 1874-1928: An Intellectual Portrait* (New York: The Free Press, 1967), p. 5. Staude points out that Scheler's uncle introduced him to the writings of Nietzsche and that the effect of those writings on young Scheler was so profound that "he was later to be known as 'the Catholic Nietzsche.'"

8. Friedrich Nietzsche, *The Genealogy of Morals: An Attack*, trans. Francis Golffing (Garden City, New York: Doubleday, 1956), First Essay, pp. 149–88.

9. Max Scheler, *Ressentiment*, trans. William Holdheim (New York: Free Press of Glencoe, 1961), p. 154. Cited in Staude, p. 43.

10. Scheler, *Problems of a Sociology of Knowledge*, p. 130.

11. Ibid., pp. 129–30.

12. Friedrich Nietzsche, *Thus Spoke Zarathustra: A Book for Everyone and No One*, trans. R. J. Hollingdale (Middlesex, England: Penguin Books Ltd., 1969), p. 189.

13. Gilles Deleuze, *Nietzsche and Philosophy*, trans. Hugh Tomlinson (New York: Columbia University Press, 1983), p. 2. Compare this description of genealogy with the more highly specified description in Foucault's "Nietzsche, Genealogy, History," in Michel Foucault, *Language, Counter-Memory, Practice: Selected Essays and Interviews*, trans. Donald F. Bouchard and Sherry Simon (Ithaca, New York: Cornell University Press, 1977). While it is certain that Scheler's thought would not be considered genealogical in the sense Foucault develops in this essay, my interpretation of technological culture might be so considered, at least in some respects. For instance, the perspectivism I sought to establish at the outset of this essay is one facet of genealogy, as described by Foucault. See *Language, Counter-Memory, Practice*, pp. 156–7. Other affinities between my interpretation of technical culture and genealogy could be cited, but the point here is not to make and defend the claim that what I am offering is a genealogy of convenience. Rather, it is simply that Scheler and I share the critical stance that genealogy takes when examining values.

14. Scheler sought to establish, through his phenomenological insights into moral acts, an absolute hierarchy of values. This hierarchy is written, presumably in the same manner as natural laws, upon the heart of each person. The possibility of delusion in regard to this hierarchy, however, remains open. For a brief examination of Scheler's hierarchy of values, see Kenneth W. Stikkers' "Introduction" to Scheler, *Problems of Sociology of Knowledge*, especially pp. 13–23.

15. Staude, pp. 40–41, and Eugene Kelly, *Max Scheler* Boston: Twayne Publishers, 1977), pp. 146–50.

16. Scheler probably would not have accepted this distinction. As mentioned in the preceding text, the will to control nature was seen by Scheler as an emanation of the bourgeois (i.e., common, plebeian) ethos. My distinction between the value of the leaders and the led in technical culture is not Schelerian; from Scheler's perspective, I might be seen as distinguishing the lead cows from the rest of the herd.

17. See "Traditional and Critical Theory," in Max Horkheimer, *Critical Theory: Selected Essays*, trans. Matthew J. O'Connell et al. (New York: Herder and Herder, 1972). Some of the traits shared by members of this school of thought are a concern for social justice, a critical stance toward commodity-exchange economies and the recognition of the social production of needs. While I have no complaints concerning these traits, there are others I find to be unacceptable. One of these is the preference among critical theorists for dialectical thinking, and another is a conception of freedom based on the elimination of external necessity. The first of these unacceptable traits will be taken up immediately in the text, and the second will become important later in the essay.

18. For example, Herbert Marcuse, *One-Dimensional Man: Studies in the Ideology of Advanced Industrial Society*, (Boston: Beacon Press, 1964), Chapter 6, "From Negative to Positive Thinking: Technological Rationality and the Logic of Domination."

19. William Leiss, *The Domination of Nature* (New York: George Brazziller, 1972), p. 121. See also Leiss, pp. 116–7, and Marcuse, *One-Dimensional Man*, pp. 158, 166.

20. Marcuse, *One-Dimensional Man*, p. 255. See also Leiss, p. 160, and Horkheimer, *Critical Theory*, pp. 221, 242.

21. See Horkheimer, p. 230, and Marcuse, *One-Dimensional Man*, pp. 236–37.

22. Deleuze, p. 159.

23. Similar considerations, I believe, explain the excessiveness of Nietzsche's *Genealogy of Morals*, which, it should be noted, was subtitled "An Attack."

24. See, for example, Steven Lukes, *Individualism* (New York: Harper & Row, 1973). Lukes points out "that privacy in its modern sense—that is a sphere of thought and action that should be free from

'public' interference—does constitute what is perhaps the central idea of liberalism" (p. 62).

25. Stuart Hampshire, *Freedom of the Individual* (New York: Harper & Row, 1965), p. 112.

CHAPTER 2. ARENDT, THE HOUSEHOLD, AND CONVENIENCE

1. I should at this point explain my choice for engaging Arendt rather than other, more current writers who have examined the household. Authors such as Ruth Schwartz Cowan, in *More Work for Mother: The Ironies of Household Technology from the Open Hearth to the Microwave* (New York: Basic Books, 1983), and Susan Strasser, in *Never Done: A History of American Housework* (New York: Pantheon Books, 1982), have done a fantastic job of exploring the differences between the introduction of technological improvements in industry and the introduction of such devices in the household. Cowan, in particular, has deflated the belief that technology has actually reduced the time and toil required of women who labor in the household. She points out how it was actually men who saved time through the industrialization of the household, as traditional male chores, such as the pounding of grain and the hauling of water, were replaced by technological improvements.

Such arguments are extremely valuable, and I will in fact borrow some information provided by these writers when I examine the consumption practices of the United States. However, most of these writers do not seriously question the value of convenience, but instead point out how that value was only incompletely developed or disseminated in the household. Since my goal is to challenge that very value, these texts are of only limited help in developing my perspective.

In a sense, these texts stand very close to the dialectical approach to technology which I am trying to avoid. That is, the concern of Strasser and Cowan is to uncover the reasons industrialization or technological improvement was limited in the household, to identify the conditions that thwarted the complete transformation of the household along the lines of convenience. What they would like to see, I assume, is the full development of that value in the realm of the household. My concern, once again, is to question that very value. To do this, however, I need not undermine the arguments presented by such writers. There is indeed much truth to their claims; it is just that the claim that I want to make is quite different. And as should become apparent, Arendt's interpretation of modernity and the distinctions she draws between the ancient and the modern are much more helpful in laying out this critique of convenience.

2. The productive activity of the ancient Greek household was not limited to activity that occurred within the dwelling of the family. Including as it did the activity of the family's slaves, the household was "the unit and the instrument of economic production" in both agriculture and industry. Will Durant, *The Life of Greece* (New York: Simon and Schuster, 1939), p. 307. For the Greeks used their slaves not only for manual labor, but also for clerical and executive work in industry, finance, and commerce. Durant, p. 279. All this activity, therefore, was considered part of the household. See also Michel Foucault, *The Use of Pleasure: The History of Sexuality, Volume Two*, trans. Robert Hurley (New York: Pantheon Books, 1985), p. 153.

3. Aristotle, *The Politics*, trans. T. A. Sinclair (New York: Penguin Books, 1962), pp. 28–29. On these pages Aristotle first suggests that it is the capacity for speech, as distinguished from the voice of animals, that sets men apart. But then he points out that the real difference between men and other animals is that men are able to perceive and arrive at a common understanding of right and wrong, justice and injustice.

4. Hannah Arendt, *The Human Condition* (Chicago: The University of Chicago Press, 1958), p. 38.

5. Ibid.

6. Ibid., p. 46.

7. Ibid.

8. Ibid., p. 68. Arendt argues that the social realm emerged with the sixteenth-century idea that the political authority of the monarch should protect and promote what until then had been considered the private interests of men.

9. Arendt discusses the classical Greek attitude toward mortality and immortality in Chapter 3 of *The Human Condition*.

10. Arendt, p. 60.

11. Ibid., pp. 17–18.

12. Ibid., p. 15. See also p. 85.

13. Ibid., p. 21.

14. Ibid., p. 314.

15. Ibid., p. 315.

16. Ibid., p. 316.

17. Ibid.

18. Further support for Arendt's claim can be found in the Catholic Church's prohibition of abortion and its recent opposition to the reproductive techniques that have been developed by medical science, not to mention the so-called suicide machine. Because life is sacred, argues the church, it should not be artificially created or ended.

19. Arendt, p. 314.

20. Ibid., p. 47.

21. Ibid., pp. 73–78. In these pages, which comprise the chapter entitled "The Location of Human Activities," Arendt argues that certain human activities, by the nature of those activities themselves, determine whether they should take place in public or in private. As she puts it, "If we look at these things, regardless of where we find them in any given civilization, we shall see that each human activity points to its proper location in the world" (p. 73).

22. Aristotle, pp. 33–34.

23. Arendt, p. 60.

24. Arendt explicitly rejects the claims made by some interpreters of Christianity that labor was glorified in the New Testament. According to Arendt, "there are no indications of the modern glorification of laboring in the New Testament or in other pre-modern Christian writers" (Ibid., p. 316).

25. Ibid., p. 34–35. Arendt points out that throughout the medieval period the household served as the model for "all human relationships," including political ones.

26. Ibid., p. 72.

27. See, for example, Jean Bethke Elshtain, *Meditations on Modern Political Thought: Masculine/Feminine Themes from Luther to Arendt* (New York: Praeger Publishers, 1986), p. 110, and Elisabeth Young-Bruehl, "From the Pariah's Point of View: Reflections on Hannah Arendt's Life and Work," in Melvyn A. Hill, ed., *Hannah Arendt: The Recovery of the Public World* (New York : St. Martin's Press, 1979), pp. 24–25.

I should point out that neither of these examples explicitly refutes my claim that Arendt held the body in contempt. In fact, in an earlier text, Elshtain made a very similar, if not the same, claim. See Jean Bethke Elshtain, *Public Man, Private Woman: Women in Social and Political Thought* (Princeton, New Jersey: Princeton University Press, 1981), pp. 56–57. It seems to me, however, that one might try to defend Arendt against such a claim by reference to her concept of natality. In discussing this concept at this point, I am trying to preempt such a defense.

28. Arendt, p. 247.

29. Ibid., p. 246.

30. See Elaine Pagels, *Adam, Eve, and the Serpent* (New York: Random House, 1988), pp. 57–77.

31. Arendt, p. 107, n. 53.

32. After creating him, "the LORD God took the man and put him in the garden of Eden to cultivate it and keep it." Genesis 2:15. In making this point, Arendt notes that, "According to Genesis, man (adam) had been created to take care and watch over the soil (adamah), as even his name, the masculine form of 'soil' indicates" (p. 107, note 53).

33. Ibid.

34. Arendt, p. 107.

35. These facts concerning the life of Christ, along with his resurrection from the dead and ascension into heaven, are essential to Christianity, and form part of the Christian creed. This creed was originally memorized by believers, and the ability to recite it would gain one's entry into an unfamiliar church. The creed is still recited as part of the Catholic mass.

36. Augustine, *The City of God*, abridged version, ed. Vernon J. Bourke, trans. Gerald G. Walsh, S.J, et al. (Garden City, New York: Image Books, 1958), p. 275. For a fuller statement of Augustine's interpretation of man's fall and resultant mortality, see *The City of God*, Book XIII, Chapter 1.

37. See note 16.

38. Augustine, p. 530.

39. See note 2.

40. These examples are borrowed from Michael H. Best and William E. Connolly, *The Politicized Economy*, second edition (Lexington, Mass: D.C. Heath and Company, 1982), pp. 54–59. Best and Connolly describe the influence that social changes have on luxuries and needs as the "social infrastructure of consumption" (pp. 56–57).

41. Herbert Marcuse, *An Essay on Liberation* (Boston: Beacon Press, 1969), p. 10, note 1.

42. This and the following citations to Xenophon's *Oeconomicus* refer to the translation of Carnes Lord, which is included in Leo Strauss, *Xenophon's Socratic Discourse: An Interpretation of the Oeconomicus* (Ithaca, N.Y.: Cornell University Press, 1970). This particular quote is from pp. 32–3.

43. Ibid., p. 38.

44. Ibid., p. 34.

45. Ibid., pp. 34–35.

46. Jean Hatzfeld, *History of Ancient Greece,* trans. A. C. Harrison (New York: W. W. Norton & Company, Inc., 1968), p. 121. Although Hatzfeld emphasizes the effects of developments in commerce and industry in ancient Greece, he does acknowledge that a large part of the population was engaged in agriculture.

In regard to the agricultural character of classical Athens, see also Carl Roebuck, *The World of Ancient Times* (New York: Charles Scribner's Sons, 1966), pp. 264–65, and R. J. Hopper, *Trade and Industry in Classical Greece* (London: Thames and Hudson Ltd., 1979), pp. 147–50.

47. Durant, pp. 269–70. See also Hopper, p. 153.

48. Roebuck, p. 265. See also Durant, p. 269, and Hatzfeld, pp. 121–22.

49. Durant, p. 268, and Hopper, p. 151.

50. Strauss, *Xenophon's Socratic Discourse*, p. 17.

51. Ibid., p. 23.

52. Arendt, p. 33.

53. Strauss, *Xenophon's Socratic Discourse*, p. 17.

54. Ibid.

55. Roebuck, p. 265.

56. Durant, p. 272, and Hopper, p. 140.

57. For a discussion of the crafts and trades of classical Greece, see Hopper, Chs. IV and VII.

58. Hopper, pp. 96–97.

59. Durant, p. 269, and Hopper, p. 93.

60. Durant, p. 269.

61. Ibid., and Hopper, p. 71.

62. For an extensive discussion of the role which the importation of grain played in Athens' naval empire, see Hopper, Chapter III, "The Import Trade—Principally Corn."

63. Roebuck, pp. 293–94, Hopper, p. 80, Durant, pp. 450–51, and Hatzfeld, p. 170.

64. Jean Hatzfeld claims that the class structure of Athens was based on agricultural distinctions. For instance, "the top category of citizens included anyone who could harvest five hundred bushels of wheat" (Hatzfeld, p. 46).

65. Writers such as Cowan and Strasser (see note 1) will surely take exception to this claim. Indeed, their research indicates that the amount of time that women spend performing household routines has not really been reduced by technological development, due to the greater frequency of the performances of such tasks as the cleaning of clothes. Besides this, there has occurred the creation of certain new tasks which accompanied some technological developments. For example, Cowan points out that indoor plumbing and toilet facilities created the task of cleaning the bathroom, and the automobile created several delivery and pick-up tasks for women that had been performed in the past by business delivery services or males of the household.

In defense of my claim, I must emphasize that when I write that members of the modern household spend less time satisfying the demands of the body than did members of the ancient household, I am using "demands" in distinction from limits of the body. I realize that people today spend a great deal of time satisfying needs in the household, but from my perspective, many of these needs are limits of the body, not demands. I am not claiming, in contradiction to the arguments of Cowan and others, that modern individuals have been freed from necessity. Quite the contrary, my claim is that the proliferation of bodily limits has resulted in an expanded realm of necessity.

66. By reproductive tasks, I mean that through the performance of such tasks individual products such as food, clothing, and even (somewhat crassly) children, are produced again or anew. I am not using the term in the more general, Marxist sense, which refers to the reproduction of the general supply of labor-power.

67. Even the reproduction of children can now be carried out in the technical production process of the laboratory. But although children are increasingly referred to as 'our most valuable resource,' they have not yet become consumer items. However, raising and caring for children has increasingly become a service to be consumed, not provided, by the household. I have in mind here not only out-of-the-home daycare, but also the television shows and videotapes which are produced specifically to entertain and instruct children.

68. Studs Terkel, *Working: People Talk About What They Do All Day and How They Feel About What They Do* (New York: Pantheon Books, 1972). Many of the white- and blue-collar workers Terkel interviewed for this book felt not only that they were severely constrained in their work life, but they had actually become part of the production machinery. "I am a robot" was a complaint often heard by Terkel (pp. xi–xii).

69. In the *Oeconomicus*, Socrates begins his interrogation of Ischomachus, an exemplary husband, with the question, "how do you spend your time and what do you do?" (Strauss, *Xenophon's Socratic Discourse*, p. 28.)

70. In "Paradigms Lost: Classical Athenian Politics in Modern Myth," a paper delivered at the 1987 Annual Meeting of the American Political Science Association, Blair Campbell argues that many modern scholars (most notably Arendt) have mythologized Greek political life in their attempt to counter liberalism's overriding concern with privacy.

71. Lewis Mumford finds in the construction of the pyramids the archetypical form of the "megamachine." See Lewis Mumford, *The Myth of the Machine: Technics and Human Development* (New York: Harcourt, Brace, Jovanovich, 1967), pp. 194–98.

72. A timely example of a political, not spatial, impediment to movement was the Iran-Iraq War, in which attacs on commercial tankers hindered the 'free flow' of oil through the Persian Gulf. The threat in this case came not just from the immediate impedance of the movement of oil, or from the hindrance of the freedom of the seas; this war also threatened all the movement which is generated by that supply

of oil. The extent of this threat is indicated by the fact that even nations that do not directly depend on oil from the Gulf were literally up in arms to ensure that the flow was not impeded.

There are, of course, many other explanations for the degree of concern about this war. Viewing it as a war to protect movement is a particularly narrow perspective, I admit. But viewing it as such is helpful for this essay, because it illustrates the distinction I am making between ancient and modern necessity. Contrast Athens, which fought to maintain the flow of food through the Hellespont, with the United States, which is willing to fight to maintain the flow of oil from the Persian Gulf. The Greeks were concerned with the bodily demand for food, while moderns are concerned with the need for movement, for which oil is essential.

73. Paul Virilio makes a similar claim about the importance of speed in modernity, although his concern lies primarily with military developments, not household consumption. I will examine some of Virilio's insights in Chapter 5.

74. This discussion of the etymology of convenience is based on information found in *The Compact Edition of the Oxford English Dictionary* (London: Oxford University Press, 1971, 1984).

75. This discussion is also based on information found in *The Compact Edition of the Oxford English Dictionary*, although I first became aware of the change in the meaning of comfort by reading Siegfried Giedion, *Mechanization Takes Command: A Contribution to Anonymous History* (New York: W. W. Norton & Company, 1948, 1969). For Giedion's brief discussion of the meaning of comfort, see p. 260.

76. Arendt, p. 2.

CHAPTER 3. MARXIST PERSPECTIVES ON CONSUMPTION

1. See Marx's discussion of commodity fetishism in *Capital: A Critique of Political Economy*, Volume 1, trans. Samuel Moore and Edward Aveling, rev. by Ernest Untermann (New York: The Modern Library, 1906), Chapter 1, Section 4, esp. pp. 81–86.

2. I am specifically referring here to Edmond Preteceille, Jean-Pierre Terrail, Michel Aglietta, and Ernest Mandel. In this chapter, I will examine some ideas from each of these authors.

3. Edmond Preteceille and Jean-Pierre Terrail, *Capitalism, Consumption and Needs*, trans. Sarah Matthews (Oxford: Basil Blackwell, 1985), p. 6. See also pp. 68–69.

4. Ibid., p. 1.

5. Ibid.

6. This is a major theme of the second essay of the text, entitled "The Historical and Social Nature of Needs," which was written by Terrail. In particular, see pp. 47–48.

7. Preteceille and Terrail note that the "very concept of consumption as a particular moment in social life came into being at almost the same time as political economy," and that the "representation of consumption as a specific autonomous practice gained precision with the historical development of capitalism." Preteceille and Terrail, p. 6.

8. Ibid., pp. 8–15.

9. Ibid., p. 7.

10. Ibid., p. 18.

11. Ibid., p. 39. By their insistence on the ultimately determinant character of the production process Preteceille and Terrail believe they have moved beyond distinctions such as natural and social, or real and artificial needs, which I discussed in Chapter 2, and rendered any controversy about such distinctions meaningless (p. 40). From Preteceille and Terrail's point of view, grounding all needs in the process of production obviates the need for such controversial distinctions, because all needs are socially produced needs. It will be recalled that in Chapter 2 of this essay, I tried to strike out a different route around such dichotomies, one which centered on the body, not the production process, and that I introduced the demand/limit distinction as an alternative.

12. Ibid., p. 41.

13. In the introductory notebook of the *Grundrisse*, trans. Martin Nicolaus (New York: Vintage Books, 1973), Marx points out the historical ignorance of bourgeois political economists (i.e., Smith and Ricardo), for whom the unencumbered individual is recognized "not as a historic result but as history's point of departure. As the Natural Individual appropriate to their notion of human nature, not arising historically, but posited by nature" (p. 83).

In the same notebook, Marx points out the various ways in which production and consumption appear to be identical, but then he rejects the Hegelian move of positing them as identical. Although production and consumption may appear as identical moments of one process, "production is the real point of departure and hence also the predominant moment. Consumption as urgency, as need, is itself an intrinsic moment of productive activity. But the latter is the point of departure for realization and hence also its predominant moment;... Consumption thus appears as a moment of production" (p. 94).

Marx also emphasized that production produces "not only the object but also the manner of consumption, not only objectively but also subjectively. Production thus creates the consumer" (p. 92).

14. Preteceille and Terrail, p. 4.

15. Since I have already belabored this point about the primacy of production, it is perhaps best if I carry on this harangue in a note. At the end of their discussion of the danger of private consumption and an examination of the sociological attempt to understand the nature of that consumption as the play of signs, Preteceille and Terrail conclude that a complete understanding of any such symbolic significance in consumption must include "a recognition of *the primacy of* relations of production" (p. 71, my emphasis). Aside from the ideal of a "complete understanding," what bothers me about this conclusion is the emphasized phrase. I would agree with the statement that an understanding of the specificity of consumption relations "must include a recognition of relations of production," but when the phrase in question is added, I must part company. The insistence on the primacy, or ultimate determinacy, of production, from my perspective, limits an understanding of consumption relations. I will soon become more specific about this claim and try to show what is missed through this insistence.

16. Louis Althusser, *For Marx*, trans. Ben Brewster (New York: Pantheon Books, 1969), pp. 200–16. Although Preteceille and Terrail concede that there is some usefulness to the antihumanistic, or structuralist, developments among some Marxists, they are critical of those Althusserians who invoke the ultimate determinant, production, only as a sort of ritual. To retain the usefulness of Althusser's work, the "concrete content" of those determinant factors must be worked out theoretically (Preteceille and Terrail, p. 84). In this sense, one can read Preteceille and Terrail's analysis of needs and consumption as the further concretization of Althusser's notion of a structure in dominance, although it is tempting to call it a 'fleshing out' instead, since they are headed toward the notion of a subject. Preteceille and Terrail, of course, would reject this bodily metaphor in favor of the structural one. In any case, their work can be seen as an elaboration of Althusser's

notion that in a complexly structured whole, "determination in the last instance by the economy is exercised precisely in the permutations of the principal role between the economy, politics, theory, etc." (Althusser, p. 213). For the purposes of the present discussion, consumption, as an element of "etc.", is shown by Preteceille and Terrail to affect the exercise of the ultimate determinant, the mode of production.

I should point out that the notion of a complexly structured whole in which the relations between the various elements change through time is also intimated by Marx in the introductory notebook of the *Grundrisse*. In a passage that focuses much more clearly on consumption than does the preceding quote from Althusser, Marx wrote:

> The conclusion we reach is not that production, distribution, exchange and consumption are identical, but that they all form the members of a totality, distinctions within a unity. Production predominates not only over itself, in the antithetical definition of production, but over the other moments as well.... A definite production thus determines a definite consumption, distribution and exchange as well as definite *relations between these different moments*. Admittedly, however, *in its one-sided form*, production is itself determined by the other moments (p. 99).

17. Preteceille and Terrail, p. 39.

18. Ibid., p. 48.

19. Ibid., pp. 56–58. While Preteceille and Terrail's account of the emergence of the need for restrictive hours legislation is based on Marx's analysis of workers' struggles during Europe's industrial revolution, their account is quite sketchy when compared to Marx's analysis in *Capital*, Volume 1. My discussion of the Factory Acts is based on *Capital*, Volume 1, Chapters X and XV, as well as Preteceille and Terrail's text.

20. Marx, *Capital*, Volume 1, p. 442. Another reason the introduction of 'labor-saving' machinery resulted in an extension of the working day is that the capitalists who first bought the machinery sought to maximize the advantage they had over firms that had not yet mechanized. At this early stage in the mechanization of a given craft, prices remained relatively high, and those who were producing more cheaply with the new machines were able to turn this differential into increased profits. The more they produced at this initial stage of mechanization, the more they could exploit this differential. See Marx, pp. 443–44.

21. Marx, p. 297, note 1.

22. Preteceille and Terrail, p. 57.

23. For a fuller explanation of the way in which the Factory Acts required an intensification and technical improvement of the forces of production, see Marx, pp. 514–26.

24. Preteceille and Terrail, p. 78.

25. Ibid., pp. 98–100.

26. Ibid., p. 123.

27. Ibid., p. 176.

28. Ibid., pp. 68–69.

29. Ibid., pp. 125–27.

30. Ibid., p. 69.

31. Ibid., pp. 108–9.

32. Ibid., p. 70.

33. Michel Aglietta, *A Theory of Capitalist Regulation: The US Experience*, trans. David Fernbach (London: NLB, 1979; original French edition, 1976), p. 23. It is ironic that Aglietta begins his explanation for this choice of the United States as a model by claiming that it is "motivated by reasons of convenience as well as principle" (p. 22).

34. Ibid., p. 23.

35. Ibid., p. 158.

36. Ibid., p. 160.

37. Ibid., pp. 130–35.

38. For a fuller discussion of the way in which craft skills and knowledge were used to resist production increases, see Harry Braverman, *Monopoly Capital: The Degradation of Work in the Twentieth Century* (New York: Monthly Review Press, 1974), pp. 97–103.

39. For a good discussion of the principles of Taylorism, see Braverman, Chapter 4 "Scientific Management," pp. 85–123. See also Michael H. Best and William E. Connolly, *The Politicized Economy*, 2nd edition (Lexington, Mass: D.C. Heath and Company, 1982), pp. 121–26. Best and Connolly point out earlier in their text how management's forcible removal of craft skills and knowledge from workers undermines certain justifications for economic inequality. See pp. 65–71.

40. Aglietta, pp. 118–9.

41. Antonio Gramsci, *Selections from the Prison Notebooks of Antonio Gramsci*, ed. and trans. Quintin Hoare and Geoffrey Nowell Smith (New York: International Publishers, 1971), p. 303.

42. Aglietta, pp. 158–59. In regard to recent uses of the concept of Fordism, see John Bellamy Foster, "The Fetish of Fordism," *Monthly Review*, 39, 10, 14 (March 1988). In this article, Foster criticizes recent leftist writers such as Michael Harrington (*The Next Left* [New York: Henry Holt, 1986]) who tend to mythologize the person of Henry Ford. He also criticizes writers such as Aglietta who avoid this mythologizing tendency, but nonetheless emphasize developments in consumption at the expense of any serious examination of the investment patterns of late capitalism. From Foster's perspective, the most significant element of Fordism lies not in the impetus it provided for consumption, but in the advances it made in regulating and disciplining the labor force outside of the factory. In other words, Gramsci came closer than Aglietta to recognizing the importance of Fordism. Foster does a good job of elaborating this disciplinary, or prohibitory, dimension of Fordism. See his discussion of Ford's "Sociological Department," which was comprised of a group of inspectors who evaluated the home life of Ford's employees. Much of the highly publicized "$5 a day" that Ford offered his workers depended upon the quality of their home life. If workers were judged to fall short of Ford's standards, a significant portion of their "high wages" was withheld from their pay. See pp. 18–20.

43. Aglietta, pp. 100–110.

44. Ibid., p. 107.

45. For a detailed examination of this cyclical process, which emphasizes the role of technological innovations in production, see Ernest Mandel, *Long Waves of Capitalist Development: The Marxist Interpretation* (London: Cambridge University Press, 1980).

46. Aglietta, p. 108. In support of his claim concerning the effect of the social norm of consumption, Aglietta presents data that indicate that "since 1946 there have not been any phases of deep depression in the formation of capital" (pp. 106–110).

47. Ibid., p. 159.

48. Ibid., pp. 179–86.

49. Ibid., p. 166.

50. Ibid. Aglietta presents data that indicate direct wage costs have risen less than indirect wages ever since the end of the Second World War, and that there was an explosion of indirect wages costs since 1965 (pp. 163–65). These indirect wage increases reflect the rise in cost of collective consumption.

51. Ibid., pp. 162–63. It is interesting to note that one of the unorganized forms of this resistance is a high and unpredictable rate of absenteeism. See Aglietta, pp. 120–21. Earlier it was noted that in the nineteenth century workers were able to resist the attempts to heighten the pace of production because these workers retained certain skills and knowledge of their craft. In the de-skilled work environment of Fordism, workers have lost this ability to resist on the floor of the factory. But now they can resist by staying away from the workplace altogether, while avoiding any pattern in doing so. In the highly synchronized production process of the twentieth century, unpredictable absenteeism is an effective hindrance to management's attempts to increase productivity and profits.

It is also interesting that management and owners look for the sources of this absenteeism in the urine of workers, rather than in the conditions of work itself. The prohibitory tendencies of Fordism, which Gramsci noted, are flourishing today in the hysteria concerning the use of drugs.

52. Ibid., p. 167.

53. Ibid., p. 165.

54. See note 4.

55. Aglietta, p. 159.

56. Ernest Mandel, *Late Capitalism*, trans. Joris DeBres (London: NLB, 1975; original German edition, 1972), p. 394.

57. Ibid., p. 390. The needs for increased consumption and time-saving commodities, which are mentioned in the previous quote, make up one of the six "main sources" of the diversification of consumption. These sources, however, appear to be overshadowed by another source—*the* main source—to which I will immediately turn in the essay.

58. Ibid., p. 387.

59. Ibid., pp. 387–88.

60. Ibid., p. 388.

61. Ibid., pp. 391–93. Mandel cites as one of the six sources of the differentiation of consumption (see note 57) "the increasing displacement of the proletarian family as a unit of production, and the tendency for it to be displaced even as a unit of consumption" (p. 391). In his development of this point (which, I should mention, is more extensive than that of any of the other "sources"), Mandel seems to equate the family and the household. Earlier, in the second chapter of this text, I emphasized that as I was using the term "household," it was not restricted to the traditional family or any other structure. What distinguishes a household for me is precisely that it is the locus of consumption. So while I have no quarrel with Mandel's assertion that the family, or the household, is no longer the site of productive activity, I can not accept his claim that the family/household is also being displaced as a unit of consumption. When labor which had been performed within a household is replaced by the use of a commodity or a service from outside the household, this does not indicate, as Mandel maintains, that "the material basis of the individual family (or household) disappears in the sphere of consumption as well" (p. 391), my parenthesis. On the contrary, I would interpret such replacements as an increase in the consumptive activity of the household. For me, it makes no difference whether these commodities are produced, or services provided, from outside the household; they are consumed on the basis of the resources of the household. To the extent that such services and commodities are provided gratis by the state, then I would agree that the household's status as a unit of consumption would be displaced, but that is not the point Mandel is making.

I should also mention that Preteceille and Terrail, despite their emphasis on the public satisfaction of social needs, concur with me on this point; they maintain that the household remains the concrete unit of consumption. See Preteceille and Terrail, p. 74. And Aglietta can also be enlisted in support of my position. See Aglietta, p. 159.

Despite this criticism, Mandel does make some important points about the changing status of the family. As women abandon the role of housewife and join the labor force as wage-earners, capital loses a source of unpaid labor which was essential to the reproduction of labor power. But capital gains more than it loses in this situation. As women work outside the home, the consumption of commodities and services provided by capital must increase in order for labor power to be reproduced. Furthermore, as wage-earners, these women add to the surplus labor which capital can accumulate (see Mandel, *Late Capitalism,* pp. 392–93). The fact that women are paid significantly less than men is another benefit which accrues to capital from the changing status of the traditional family. For a discussion of women's wages as "supplementary wages," see Aglietta, pp. 171–73.

62. Mandel, *Late Capitalism,* pp. 390–91.

63. Ibid., p. 378.

64. Ibid., p. 381.

65. Aglietta, p. 160. Again, I should mention that this claim of Aglietta's is seriously challenged by recent examinations of household technology. See Chapter 2, note 1.

66. Mandel, *Late Capitalism,* pp. 378–79.

67. Aglietta, p. 158.

CHAPTER 4. SETTLING AMERICAN SPACE

1. Michel Aglietta, *A Theory of Capitalist Regulation: The US Experience,* trans. David Fernbach (London: NLB, 1979; original French edition, 1976), p. 74.

2. Ibid., p. 73.

3. Ibid., p. 31. Aglietta cites this section of *Capital* in support of his claim that labor-power is not a commodity, as Marx mistakenly assumed in other places in his writing.

4. Karl Marx, *Capital: A Critique of Political Economy,* Volume 1, trans. Samuel Moore and Edward Aveling, rev. by Ernest Untermann (New York: The Modern Library, 1906), p. 787.

5. Ibid.

6. E. C. K. Gonner, *Common Land and Enclosure* (New York: Augustus M. Kelley, Bookseller, 1966; first published, 1912), p. 48; and J. L. Hammond and Barbara Hammond, *The Village Labourer* (London: Longman Group Limited, 1978; first published, 1911), pp. 6–7. These two texts present very different interpretations of the enclosure movement in England. The Hammonds' text is close to the position Marx held on enclosure, and presents the enclosures as violent uprootings of peasant farmers. Gonner, on the other hand, presents the enclosure movement as a benefit to the agricultural and manufacturing activity of England, which had very little adverse effect upon the peasantry. For my purpose, which is to contrast the English and American experiences of land or space, both of these texts are helpful, despite their significant differences.

7. The preceding description of common rights is based on Gonner, pp. 3–36.

8. Ibid., p. 82.

9. Hammond and Hammond, pp. 58–61. See also Gonner, p. 362.

10. Mr. Bishton, in the *Report on Shropshire*, 1794, as quoted in Hammond and Hammond, p. 9.

11. Marx, *Capital*, Volume 1, p. 838, note 1.

12. Ibid., p. 838. Although Marx does not explicitly refer to him in this chapter, John Locke is an excellent example of such a confused political economist. Locke emphasizes that a person's ability to appropriate as his or her own some part of that which God had given to all in common depends not on the consent of others, but on the labor expended on that property. To quote a famous passage from the *Second Treatise*:

> Though the earth and all inferior creatures be common to all men, yet every man has a 'property' in his own 'person.' This nobody has a right to but himself. The 'labour' of his body and the 'work' of his hands, we may say, are properly his. Whatsoever, then, he removes out of the state that Nature hath provided and left it in, he hath mixed his labour with it, and joined to it something that is his own, and thereby makes it his property (John Locke, *Of Civil Government: Two Treatises* [London: J. M. Dent & Sons Ltd., 1924], p. 130).

Shortly after this statement of the foundation of private property, Locke cites an example of the creation of private property which reveals his confusion. "Then the grass my horse has bit, the turfs *my servant has cut*, and the ore I have digged in any place, where I have a right to them in common with others, become my property..." (Ibid., my emphasis). If the servant, who presumably fits the description of a person found in the long quote above, does the cutting of the turf, how can the master of this servant claim these turfs as his property? How does Locke get around his injunction that a person has an exclusive right to his or her body and the labor of that body?

C. B. Macpherson argues that for Locke there was no contradiction between these two statements. Rather, Locke was universalizing the conditions of seventeenth-century England, where the absence of unsettled land and the enclosure of the commons made wage-labor a necessity for those without land. "For Locke, then, a commercial economy in which all the land is appropriated implied the existence of wage-labour. And since Locke was reading back into the state of nature the market relations of a developed commercial economy, the presumption is that he read back the relation along with other market relations." C. B. Macpherson, *The Political Theory of Possessive Individualism: Hobbes to Locke* (Oxford: Oxford University Press, 1962), p. 217.

For a more detailed and insightful reading of Locke's theory of property, see Macpherson, pp. 194–221. And for a critique of Macpherson's interpretation of Locke as an apologist for capitalist appropriation, see Karen Iversen Vaugh, *John Locke: Economist and Social Scientist* (Chicago: University of Chicago Press, 1980), pp. 100–107.

13. Marx, *Capital*, Volume 1, p. 838.

14. See the first quote from Locke's *Second Treatise* in note 12.

15. Marx, *Capital*, Volume 1, p. 843.

16. Ibid.

17. Ibid., pp. 846–47.

18. Ibid., p. 847.

19. Ibid., p. 797. For brief descriptions of laws of tillage, see also Marx, p. 790, and Gonner, p. 344. See Hammond and Hammond, p. 102ff. for a description of the "allotment movement," which sought to have agricultural land set aside for the use of the poor.

20. Everett Dick, *The Lure of the Land: A Social History of the Public Lands from the Articles of Confederation to the New Deal*, (Lincoln: University of Nebraska Press, 1970), pp. 6–7; Aaron M. Sakolski, *Land Tenure and Land Taxation in America* (New York : Robert Schalkenbach Foundation, Inc., 1957), p. 69; and Richard A. Bartlett, *The New Country: A Social History of the American Frontier: 1776–1890* (New York: Oxford University Press, 1974), pp. 66–67.

21. Dick, pp. 5–6; and Bartlett, p. 25.

22. Dick, p. 50.

23. Dick, pp. 3–5; and Sakolski, pp. 41–43.

24. Sakolski, pp. 33–42; and Roy M. Robbins, *Our Landed Heritage: The Public Domain: 1776–1870* (Lincoln: University of Nebraska Press, 1976; original edition, 1942), pp. 6–8.

25. New York ceded its western lands to the Continental Congress in 1780, and Virginia gave up most of its territory north of the Ohio River in 1784. By 1802, the remaining five states with western lands had followed suit. See Robbins, pp. 4–5.

26. Dick, pp. 20, 50, 55; and Robbins, pp. 17–8.

27. Dick, p. 50; Sakolski, p. 129; and Seymour Dunbar, *A History of Travel in America* (Indianapolis: The Bobbs-Merrill Company, 1915), p. 82.

28. Dick, pp. 50–54; and Bartlett, p. 28.

29. Dick, pp. 53–54; Bartlett, pp. 25–28; and Robbins, pp. 227, 234. The military campaigns against the various Indian tribes, as well as the larger environment of duplicity in which relations with the Indians were carried out by the U.S. government, are beyond the scope of this essay. This should not be understood as an apology for the genocide which underpins the formation and expansion of the United States. It is just that my concern here lies with the subjugation of independent white settlers and the role that subjugation played in shaping the typically American attitude toward technology and convenience.

30. Bartlett, p. 69.

31. Bartlett, pp. 69–72; Sakolski, p. 85; Robbins, p. 8; and Dick, pp. 19–21.

32. Ibid.

33. Bartlett, p. 68–69.

34. Ibid.; Robbins, p. 7; and Dick, pp. 19–20.

35. Gonner, pp. 137, 184. Part of Gonner's thesis is that the enclosure movement did not reduce the amount of land under cultivation, as more radical interpreters have maintained, but actually increased agricultural acreage.

36. Bartlett, p. 68.

37. In his essay "The Question Concerning Technology," Martin Heidegger characterizes modernity as an age in which man attempts to set the world into a grid-like order. His notions of "Enframing" and the "World Picture" are helpful in understanding the impetus behind the American response to open, unsettled space. "Enframing means the gathering together of that setting-upon which sets upon man, i.e., challenges him forth, to reveal the real, in the mode of ordering, as standing-reserve" (Martin Heidegger, *The Question Concerning Technology and Other Essays*, trans. William Lovitt [New York: Harper & Row, Publishers, 1977], p. 20). The U.S. government policy of surveying the public domain before settlement, of setting this space in order right from the start, can be read as an example of Enframing.

In the essay, "The Age of the World Picture," Heidegger claims that "the fundamental event of the modern age is the conquest of the world as picture. The word 'picture'...now means the structured image...that is the creature of man's producing which represents and sets before. In such producing, man contends for the position in which he can be that particular being who gives the measure and draws up the guidelines for everything that is" (Ibid., p. 134). Heidegger goes on to say that as the modern age races toward its fulfillment, there appears "everywhere and in the most varied forms and disguises the gigantic" (Ibid., p. 135). The gigantic is that which is incalculable; it is an "invisible shadow which is cast around all things everywhere" as the world becomes increasingly established as a picture produced by man (Ibid.).

In regard to the land policies of the United States as an example of man's picture production, the gigantic appears as the inability of the rectilinear survey scheme to establish identical, absolutely equal sections. As the surveys moved farther north, away from the original base line, the size of the sections had to be periodically reduced to compensate for the difference between magnetic north, on the compass, and true north. A degree of incalculability emerged even in this highly ordered survey system. See Bartlett, p. 68f.

38. Bartlett, p. 72; Robbins, p. 8; and Sakolski, p. 83. Robbins mentions a 36-dollar surveying fee per township, while Bartlett describes the rate as being one dollar per section.

39. Robbins, pp. 9, 26–27.

40. Bartlett, pp. 72, 74–75; and Sakolski, p. 83.

41. Robbins, p. 9.

42. Bartlett, pp. 180–81.

43. Ibid., pp. 72–3; and Sakolski, p. 83.

44. Sakolski, pp. 83–84; and Dick, p. 8.

45. Sakolski, pp. 85–86; and Dick, p. 10.

46. Bartlett, p. 74; and Dick, p. 9.

47. Dick, pp. 9–10; Sakolski, p. 94; and Robbins, pp. 24–25.

48. Aglietta, p. 159; and Ernest Mandel, *Late Capitalism*. Trans. Joris DeBres (London: NLB, 1975; original German edition, 1972) pp. 384–85.

49. Robbins, p. 32.

50. See note 45.

51. Robbins, pp. 31–32; Sakolski, p. 124; and Dick, p. 11. Robbins claims that land sales by 1819 amounted to forty-four million dollars; Sakolski claims that the total was over forty-six million dollars. Both agree that this debt was largely unpaid.

52. Robbins, pp. 31–32; Dick, pp. 11–12; and Bartlett, pp. 74–75, 97–98.

53. Ibid.; and Sakolski, p. 84.

54. Dick, pp. 13, 57, 186, 199–201, 356–57; Sakolski, pp. 124–26; Bartlett, p. 74; and Robbins, p. 65.

55. Bartlett, p. 75; and Robbins, pp. 48–49. Robbins estimates that in the late 1820s, two-thirds of the population of Illinois was comprised of squatters. Other frontier historians estimate that in the early nineteenth century, "from one-half to two-thirds of the settlers in any new region were normally 'squatters'" (Ray Allen Billington and Martin Ridge, *Westward Expansion: A History of the American Frontier*, Fifth edition [New York: Macmillan Publishing Co., 1982; original edition, 1949], p. 336).

56. Dick, pp. 12–13, 56, 59; and Robbins, p. 66.

57. Paul W. Gates, *The Farmer's Age: Agriculture, 1815–1860*, Volume III of The Economic History of the United States (New York: Harper & Row, Publishers, 1960), p. 73.

58. Ibid.; and Dick, p. 56.

59. Robbins, p. 67. Dick claims that "these organizations were universal in the area from Indiana west to Nebraska and Kansas and north to the states bordering Canada..." Dick, p. 59. By 1840, more than one hundred claim clubs existed in Iowa. See Billington and Ridge, p. 420.

Although the claim club flourished after 1820, it had appeared as early as 1769, in opposition not to speculators, but colonial proprietors (Dick, p. 4).

60. In the decade of the 1820s, which was the period in which settlers could not buy land on government credit and had to compete at auction (after 1831, for reasons which I will get to shortly, squatters could often buy their land at the minimum price without auction), the average price per acre was 1.33 dollars, just 8 cents above the minimum (Dick, pp. 356–57; see also Robbins, pp. 65, 81).

61. Dick, pp. 62–63, 66–67.

62. Dick, pp. 66–67, 102–103.

63. Dick, p. 64; and Robbins, p. 67.

64. Dick, p. 67; Sakolski, pp. 129–30; and Robbins, p. 50.

65. This is not to say that speculators were thwarted in their accumulation of land, but just that settlers no longer had to pay 'hush money' to speculators to keep them from bidding. Actually, speculators were able to accumulate land under the Pre-emption Acts by presenting fraudulent evidence and testimony attesting that they had actually settled the land (see Bartlett, pp. 78, 240; and Dick, pp. 112, 233, 357). Members of the claim clubs also abused the Pre-emption Acts (see Dick, pp. 102–103).

66. Dick, p. 68; Sakolski, p. 130; Robbins, pp. 89–90; and Bartlett, pp. 75, 113, 240. For a detailed discussion of the passage of the Pre-emption Act of 1841, the so-called "log cabin bill," see Robbins, pp. 80–91.

67. Dick, p. 357; and Robbins, p. 89.

68. Dick, p. 139; Robbins, pp. 206–207; Sakolski, pp. 136–37; and Bartlett, p. 113.

69. Most of the authors of the various texts upon which I have relied in this discussion of land policy would not agree with my perception of irony in these developments. Robbins goes so far as to claim that the turn of events which began with the election of Andrew Jackson as president, included the Pre-emption Acts of the 1830s, and culminated with the Act of 1841 indicated that "the frontier interests at last triumphed" (Robbins, p. 89), and that "for the first time in world history, a frontier society came into control of national polity" (Ibid., p. 45). Needless to say, I do not share Robbins' enthusiasm.

Dick, on the other hand, tends to treat the apparently progressive developments in land policy as detriments to the national character. For example, of the credit system in effect from 1800 to 1820, he writes, "More serious than the amount of money involved, however, was the damage done to the citizens' standard of honor" (Dick, p. 10). Citizens regularly defaulted on their debts, a trait inimical to a capitalist economy. My criticism of the credit system, on the other hand, is that it made land available to settlers on the terms of the capitalist exchange economy.

Bartlett, in his text, points out the benefits which were provided to the country by speculators, along with the many detriments. He even calls them "harbingers of frontier advancement" (Bartlett, p. 60; see also pp. 60–62). From my perspective, speculators served the role that the government would have served according to Wakefield's plan: they drove the price of land up and made it less accessible to actual settlers.

Of the various authors I have cited, Sakolski comes closest to my perspective. He deplores the way in which speculators were allowed to undermine every government attempt to foster the sale of public land to small homesteaders (see, for example, Sakolski, p. 98). My argument goes beyond the problem of speculation, however, and the remainder of this chapter will be spent describing other forces that undermined the apparently progressive developments in U.S. land policy.

70. Dick, pp. 153–54. See also Sakolski, pp. 136–37. Such commutations were fraudulent when accomplished by speculators because the terms of the Act required the applicant to file a sworn affidavit stating that the land was "for the purpose of actual settlement and cultivation..." and was "for the exclusive use and benefit" of the applicant (Robbins, p. 207). To apply for the land under the Act with the intent to sell it for a profit was a fraud.

71. Robbins estimates that only one billion acres remained of the public domain by 1862 and that only one-third of that was arable (see Robbins, p. 236).

72. Bartlett, pp. 180–87; and Gates, p. 140f.

73. Bartlett, p. 176.

74. Bartlett, pp. 176–77; and Siegfried Giedion, *Mechanization Takes Command: A Contribution to Anonymous History* (New York: W. W. Norton & Company, Inc., 1969; original edition, 1948), p. 147.

75. Giedion, pp. 147, 149; and Bartlett, p. 177.

76. Giedion, p. 144; and Gates, p. 279.

77. Gates, p. 280. See also Bartlett, p. 199.

78. Gates, p. 281. See also Giedion, p. 149, in regard to the diversification of plows in America.

79. Gates, pp. 281–82; and Bartlett, pp. 203–204.

80. Giedion, p. 142.

81. Bartlett, p. 199; and Gates, p. 166. Bartlett calls wheat "the great cash-crop cereal," but it might better be called a commercial crop. Prior to the extensive development of the railroads, a point I will get to

later, midwestern wheat was "the basis of barter, the standard of value" (Gates, p. 166).

82. Gates, pp. 166, 169.

83. Giedion, p. 135.

84. Gates, p. 285; and Bartlett, p. 199.

85. Gates, p. 285; and Bartlett, p. 199.

86. Gates, p. 288.

87. Gates, p. 287; and Bartlett, p. 205.

88. Wakefield's lament about the labor supply in colonies (see p. 68 of this chapter) applied to farmers as well as industrialists. See Gates, pp. 271–79, for a discussion of farm labor. The problem was not just that immigrants and emigrants, like the farmers themselves, were looking to establish their own farms, but that construction and railroad jobs in the midwest paid wages much higher than the farmer could afford.

89. Giedion, p. 152; and Bartlett, p. 205. The improvement over the sickle, of course, was much greater.

90. Gates, p. 286; Giedion, p. 152; and Cyrus McCormick, *The Century of the Reaper* (New York and Boston: Houghton, Mifflin and Company, 1931), p. 36.

91. All of this information on McCormick's Chicago factories is found in McCormick, pp. 34–41.

92. Giedion, p. 153.

93. Gates, pp. 287–88; Giedion, p. 166; and McCormick, pp. 50–51. For earlier prices, see Bartlett, p. 205.

94. Gates, pp. 166–67.

95. Giedion, p. 161.

96. Ibid., p. 159.

97. Gates, p. 288.

98. Bartlett, p. 206; and Giedion, p. 145.

99. Gates, p. 288.3

100. Ruth Schwartz Cowan, in *More Work For Mother* (see Chapter 2, note 1), contrasts a typical meal prepared by an eighteenth-centu-

ry household with that prepared by a nineteenth-century household. While the earlier household's one-pot meal required the efforts of the husband, wife, and children and was comprised of foodstuffs produced by the household, Cowan emphasizes that even that earlier household was involved to some extent in the commercial market, at least for durable goods such as pots and andirons, and for some staples such as salt. Cowan, pp. 31–32. For her detailed comparison of the two centuries' households, see Cowan, pp. 16–68.

101. Gates, p. 292. In regard to commercial food production, Cowan has pointed out that by the time of the U.S. Civil War, the domestic production and consumption of grain was largely obsolete. Wheat flour had replaced corn as the dominant grain used in the household, and by 1860, commercial flour milling was the largest industry in the American economy. Cowan claims that the value of the product of the flour industry was twice as large as that of the cotton industry, and three times larger than that of the steel industry (see Cowan, pp. 47–48). Cowan's larger argument is that the commercial production of flour did not reduce the amount of time nineteenth century women spent producing breads and other baked goods. Compared to traditional cornbread, white bread (which became the standard) required more time (to let the yeasted dough rise) and more labor (to knead the bread to develop gluten). And the baked goods which could be produced from the finer wheat flours were also time-consuming delicacies which became the standard of nineteenth-century households. But if women did not benefit from the commercial production of wheat flour, men did save time on certain household chores required for the preparation of corn into meal. It had been the man's responsibility to pound the corn into meal or to drive the corn to the local mill for grinding. For Cowan's interesting discussion of the industrialization of flour milling and its consequences for the household division of labor, see Cowan, pp. 46–53.

102. McCormick, pp. 50–51. See also J. R. Dolan, *The Yankee Peddlers of Early America* (New York: Clarkson N. Potter, Inc. 1964), pp. 250–51.

103. Gates, p. 293.

104. There are notable exceptions to this trend. Amish farmers have successfully resisted the mechanization of agriculture and continue to rely primarily on human and animal labor to produce their crops. Although the scale of these Amish farms is significantly less than the enormous farms of the midwest, they have been able to remain prof-

itable even as those larger farms have been sinking deeper into debt and bankruptcy. Of course, the profits of the small Amish farms are modest as well.

CHAPTER 5. SETTING BODIES IN MOTION

1. Karl Marx, *Capital: A Critique of Political Economy*, Volume 1, trans. Samuel Moore and Edward Aveling, rev. by Ernest Untermann (New York: Modern Library, 1906), pp. 805–809.

2. Adam Smith, *An Inquiry into the Nature and Causes of the Wealth of Nations*, 2 vols. (London: Methuen & Co., 1904), Volume 1, pp. 137–42.

3. J. L. Hammond and Barbara Hammond, *The Village Labourer* (London: Longman Group Limited, 1978; first published, 1911), pp. 70–73; and Smith, pp. 137–42.

4. Hammond and Hammond, p. 202.

5. Hammond and Hammond, pp. 133, 140, 145, 208.

6. Paul Virilio, *Speed and Politics: An Essay on Dromology*, trans. Marc Polizzotti (New York: Semiotext(e), 1986; original French edition, 1977), Part One, "The Dromocratic Revolution," pp. 3–24.

7. Paul Virilio, *Pure War*, an interview conducted by Sylvere Lotringer, trans. Marc Polizzotti (New York: Semiotext(e), 1983), pp. 44–45.

8. Ibid., p. 45.

9. Virilio, *Speed and Politics*, p. 46.

10. Ibid., p. 30. In the United States, of course, this freedom was attained not through political revolution, as it was in France (it is in the context of the French Revolution that Virilio makes this claim about the obligation to move), but was one of the material conditions of America.

11. Despite his military focus, Virilio's treatment of speed engages mine in several ways. For instance, Virilio claims that "speed is the hope of the West; it is speed that supports the armies' morale. What 'makes war convenient' is transportation, the armored car, able to go over every kind of terrain. With it earth no longer exists" (Virilio, *Speed and Politics*, p. 55). Eventually I too will treat the pursuit of speed as an attempt to overcome the limits of the earth as well as the body.

12. Ibid., pp. 26–27.

13. Seymour Dunbar, *A History of Travel in America* (Indianapolis: The Bobbs-Merrill Company, 1915), pp. 14–23, 198, 222–25; and Richard A. Bartlett, *The New Country: A Social History of the American Frontier: 1776–1890* (New York: Oxford University Press, 1974), pp. 276–78, 289–90.

14. Dunbar, pp. 27–28, 226–27.

15. Ibid., pp. 31, 312.

16. Ibid., p. 312.

17. Balthasar Henry Meyer et al., *History of Transportation in the United States before 1860* (Washington: Carnegie Institute of Washington, 1917), p. 118.

18. Dunbar, pp. 321–2; and George Rogers Taylor, *The Transportation Revolution, 1815–1860*, Volume IV of The Economic History of the United States (New York: Rinehart & Company, Inc., 1951), p. 17.

19. Taylor, p. 17; Dunbar, p. 320; and Meyer et al., pp. 16–17, 52, 299. Roads made of gravel on a stone foundation were called "macadamized" roads, after Macadam, who greatly influenced road construction in England. Meyer et al. point out that Macadam learned this technique in the United States, where it first appeared. See Meyer et al., p. 52.

20. Taylor, p. 29; Meyer et al., pp. 37–50; and Dunbar, p. 640.

21. Taylor, pp. 17–18; Meyer, p. 63; and Bartlett, p. 278.

22. Bartlett, pp. 291–95; and Dunbar, pp. 201–205.

23. Meyer et al., pp. 7–12; and Dunbar, pp. 133–69.

24. Meyer et al., p. 9.

25. Taylor, pp. 19, 22; Meyer et al., p. 16; and Dunbar, pp. 692, 697. Dunbar claims that the turnpike reached Wheeling in 1817, while Meyer indicates that the road was opened to travel in 1818.

26. Dunbar, p. 692. However, the National Road never did reach St. Louis on the Mississippi River, as originally intended. It ended in Vandalea, Illinois due both to the success of railroads and the political conviction that the federal government should not be directly involved in the construction and operation of improvements in interstate travel.

See Dunbar, pp. 701–16, for an account of the political controversy surrounding the issue of federal financing of internal improvements.

27. Dunbar, p. 712. This occurred between 1829 and 1837, during Andrew Jackson's two presidential terms.

28. Dunbar, p. 717; and Meyer et al., p. 18.

29. Meyer et al., pp. 122–24.

30. And two percent of these sales was to be used by the federal government for the construction of roads. This policy helped to finance the National Road during the period in which the federal government operated it. See Meyer et al., p. 13; Taylor, p. 19; Dunbar, pp. 698–700; Everett Dick, *The Lure of the Land: A Social History of the Public Lands from the Articles of Confederation to the New Deal* (Lincoln: University of Nebraska Press, 1970), p. 160; and Roy M. Robbins, *Our Landed Heritage: The Public Domain: 1776–1870* (Lincoln: University of Nebraska Press, 1976; original edition, 1942), p. 27.

31. Henry Pickering Walker, *The Wagonmasters: High Plains Freighting from the Earliest Days of the Santa Fe Trail to 1880* (Norman, Oklahoma: University of Oklahoma Press, 1966), p. 204.

32. Taylor, p. 142. Although the actual speed of wagon trains was not much faster than that of pack trains, wagons eliminated the time-consuming daily task of loading all of one's cargo onto the animals in the morning and unloading in the evening. See Walker, p. 204.

33. Meyer et al., p. 17. Meyer is quoting from Searight's *The Old Pike*.

34. Bartlett, p. 281; and Meyer et al., pp. 68–70.

35. Dunbar, pp. 270–71.

36. Ibid., pp. 271–72, 284–86; Bartlett, pp. 307–308; Meyer et al., pp. 96–97; and Taylor, p. 56.

37. Ibid.

38. Bartlett, p. 310; Dunbar, pp. 281–82; Meyer et al., p. 98; and Taylor, p. 57.

39. Taylor, p. 57; and Bartlett, pp. 308–310.

40. Meyer et al., pp. 102–103; Bartlett, pp. 317–8; and Dunbar, pp. 389–91.

41. Meyer et al., pp. 94–95, 102.

42. Dunbar, pp. 393–94; Meyer et al., p. 104; and Taylor, p. 63.

43. Meyer et al., p. 115.

44. Dunbar, pp. 394–95; Meyer et al., p. 104; and Taylor, p. 63.

45. Taylor, pp. 63–64. See also Bartlett, pp. 319, 321.

46. Taylor, p. 64; and Bartlett, p. 321.

47. Bartlett, p. 320; and Dunbar, p. 316, note 2.

48. Taylor, p. 142.

49. Of course, where other routes were available, travelers could avoid the tolls of the turnpike (see Meyer et al., p. 55), but the most direct routes, those which had been used by generations of Indians and white men, were the ones for which turnpike charters were granted. The alternatives, where available, were considerably inferior in directness as well as condition.

"Shunpikes" also emerged, routes that circumvented particular tollgates, but resistance to the turnpike companies was very limited here too. See Bartlett, p. 281.

50. Taylor, p. 64; and Meyer et al., p. 103.

51. Taylor, pp. 32, 34–36, 52; and Dunbar, p. 770.

52. Dunbar, pp. 772–763. Dunbar cites several examples of such canals built before 1817.

53. Meyer et al., pp. 103, 280; and Taylor, pp. 67–68. The federal government invested in the private company that constructed this canal. Such investment in river improvements was the extent of the federal government's role in promoting steam navigation.

54. Dunbar, pp. 772–75, 848; and Taylor, pp. 37–43.

55. Dunbar, p. 848.

56. Taylor, p. 33; and Bartlett, pp. 312–14. Dunbar, p. 783; and Meyer et al., p. 193, claim that the canal was 363 miles long when completed. See also Meyer et al., p. 241.

57. Taylor, p. 45; Bartlett, p. 317; and Dunbar, pp. 850–51.

58. Taylor, p. 44; and Bartlett, p. 314.

59. Dunbar, pp. 794–99; Taylor, p. 44; and Bartlett, p. 314.

60. Taylor, pp. 44, 54; and Bartlett, p. 314.

61. Taylor, p. 142; and Dunbar, p. 851.

62. Dick, p. 161. See also Robbins, pp. 89–90. To ensure that the federal government would not lose money on this arrangement, the minimum price of its land in these grant areas was doubled to $2.50 per acre, making up for the revenue lost on the land given to the states while raising the value of the land which was granted.

63. The first chartered railroad was the Baltimore and Ohio Railroad, which was chartered in 1827 and opened in 1830. This was the first in the United States to carry passengers as well as freight. The vehicles on these tracks were horse drawn. See Meyer et al., p. 398; and Taylor, p. 77.

64. Dunbar, p. 1054.

65. Prior to the demonstration of the capabilities of the steam engine, not only were horses considered to be the choice motive power for railroads, but it was also thought that vehicles might be independently owned by travelers, who would pay, as on a turnpike, for the use of the thoroughfare. Small, horse-drawn vehicles were developed, as were vehicles which were powered by a horse tread. See Taylor, pp. 82–83; Meyer et al., p. 311; and Dunbar, pp. 931–32.

66. Aaron M. Sakolski, *Land Tenure and Land Taxation in America* (New York: Robert Schalkenbach Foundation, Inc., 1957), pp. 160–61; and Dick, p. 162. Taylor claims that the land granted to the states crossed by the Illinois Central amounted to more than 3.7 million acres (p. 96).

67. Taylor, p. 96.

68. Sakolski, pp. 165–66; and Bartlett, pp. 327–8.

69. Dick, p. 165; and Sakolski, p. 167.

70. Sakolski, pp. 166, 173; and Dick, p. 164.

71. Bartlett, pp. 332–34; Dick, pp. 174–75; and Sakolski, p. 172.

72. Bartlett pp. 33–34; and Dick, p. 174. See note 62 as well.

73. Sakolski, p. 166; Dick, p. 176; and Bartlett, p. 327.

74. Dick, p. 176; and Bartlett, p. 327.

75. It was not until 1877, in *Munn v. Illinois* that the Supreme Court of the United States recognized the validity of such state regulations of grain elevators.

76. Dick, p. 162.

77. Ibid., pp. 165–66.

78. Ibid., pp. 169–70.

79. See Ruth Schwartz Cowan, *More Work for Mother: The Ironies of Household Technology from the Open Hearth to the Microwave* (New York: Basic Books, 1983), pp. 79–85. Cowan also points out that the transportation system which developed around the automobile provides a different model of household industrialization than other technological systems of the household, such as clothing, food, and health care systems. In those other areas, housewives went from being the producers of certain goods and services to the consumers of such things. But in the realm of conveyance or movement, housewives had traditionally been the consumers of transportation service, but were rendered producers of transportation through the family-owned automobile.

80. Ivan Illich, *Toward a History of Needs* (New York: Pantheon Books, 1977), p. 121.

81. Ibid., p. 122.

82. Cowan, pp. 82–83. Cowan writes that "the speed with which the automobile became an integral part of daily life in America was, in historical terms, astounding."

CHAPTER 6. WEBER, PROTESTANTISM, AND CONSUMPTION

1. Anthony Giddens makes this point in his "Introduction" to the text. Giddens also categorizes the various criticisms that have been made of Weber's argument. See Max Weber, *The Protestant Ethic and the Spirit of Capitalism*, trans. Talcott Parsons (London: Unwin Paperbacks, 1987), pp. xix–xxvi.

The debate surrounding Weber's text still continues. In a fairly recent article published in the journal *Telos*, Luciano Pellicani presents a litany of counter-examples which are intended to give the lie to Weber's claim that Calvinist Puritanism was essential to the development of the rational, calculative mindset of the modern entrepreneur (Luciano Pellicani, "Weber and the Myth of Calvinism," *Telos*, 75, Spring 1988, pp. 57–85). Generally, Pellicani argues that the entrepreneurial spirit was present long before Calvinism emerged and that Weber's argument blinds scholars to those earlier manifestations of that spirit. He concludes his article by saying that *The Protestant Ethic* is "a work which should be put aside once and for all" (Ibid., p. 85). A few months later, *Telos* dedicated an entire issue to Weber, and several writers, apparently still deluded by the myth of Calvinism, took Pellicani to task for his harsh dismissal of Weber's argument. See *Telos*, 78, Winter 1988–89.

It seems that the lines that have been drawn in the debate resolve to two different ways of reading Weber. Critics of Weber, particularly Pellicani, seem to take Weber's argument much more seriously, or dogmatically, than Weber himself was wont to do. They seem to read Weber's argument as implying that capitalism could never have developed without the psychological transformations which were accomplished by Calvinism. The defenders of Weber, on the other hand, offer a much more generous, or perhaps relaxed reading of his argument. Calvinism, in this reading, was a contributing factor, and a powerful one, in the development of capitalism. It will become apparent in the text that I think the latter approach is both fairer to Weber and more fruitful than the former.

2. Weber, p. 91.

3. Ibid., p. 190.

4. Ibid., note 84 of Chapter V, pp. 277–78.

5. Weber recognized that capitalism had existed prior to the Reformation, and that large banks and merchants were not new to the modern age, but he distinguished such capitalism as traditional. The difference between traditional and nontraditional capitalism is that under the former form, the labor power of the society or community was not organized according to rational principles of profit accumulation. For Weber, the emergence of those principles marks the boundary between the Middle Ages and the modern age. See Weber, note 23 of Chapter II, p. 200. It is with nontraditional, or modern, capitalism that Weber was concerned. It is his claim that the Protestant Reformation played an important role in the emergence of this rational form of capitalism.

6. Weber, p. 176.

7. Ibid., p. 179.

8. This term emerges only at the end of Weber's text, and it does so in only two instances. (Weber p. 180). Nevertheless, it is as important as any of the other concepts which Weber describes more fully in

The Protestant Ethic, especially in that it indicates Weber's concern for the modern capitalist order, in which religion no longer plays an important role.

9. Ibid., p. 55.

10. Ibid., note 108 of Chapter V, p. 282.

11. Ibid., p. 181.

12. Ibid., p. 84.

13. Ibid., p. 80.

14. Martin Luther, *Luther's Works*, Volume 28, "Commentaries on I Corinthians 7; I Corinthians 15; Lectures on Timothy," ed. Hilton C. Oswald (Saint Louis: Concordia Publishing House, 1973), p. 45.

15. Ibid., p. 52. This necessity/pleasure distinction of Luther's is not susceptible to the criticism of the needs/wants type distinction I offered earlier (see Chapter 2, pp. 29–31). By pleasure, Luther is not referring to something artificial or non-essential to the body; on the contrary, he has in mind powerful lusts which race through man's corrupted flesh. See, for example, Luther, *Works*, Volume 1, p. 203, where Luther claims that "the husband has a raging lust kindled by the poison of Satan in his body."

This pleasure/necessity distinction of Luther's seems to be derived from Augustine's distinction between *uti* and *frui*. In his "De doctrina Christiana," Augustine distinguished things which were to be used and things which were to be enjoyed. He identified as "the proper object of our enjoyment...the Father, Son, and Holy Ghost, the Same who are the Trinity..." *The Writings of St. Augustine*, Volume 2 of The Fathers of the Church Series (Washington: The Catholic University of America Press, 1947), p. 30. As for life in this world, Augustine taught that "we must use this world, and not enjoy it" (Ibid., p. 30) lest we become confused by false pleasures, which are fleeting and changeable, and lose sight of "that Truth which lives without change" (Ibid., p. 34). Later in this chapter, I will indicate how Luther and Calvin broke with this Augustinian tradition.

16. I Corinthians 7:17–28. The seventh chapter of the first epistle to the Corinthians is concerned with questions the newly converted Corinthians had about Christian marriage practices. Paul's advice was that it really did not matter if one was married or not, or whether one was married to a Christian or not. As long as one's marital status did not interfere with one's faith, it made no difference for one's relation to

God. When Paul expounds this idea, he mentions the examples of circumcision and slavery. None of these conditions matter.

17. Luther, *Works*, Volume 28, p. 43.

18. Weber, p. 85.

19. Ibid., p. 86.

20. Weber does indicate that this change in Luther's position on worldly activity was due, in part, to Luther's increasing involvement in worldly affairs and partly to his experience with peasant rebelliousness (Weber, pp. 84–85). But Weber mentions these sources of Luther's traditionalism only in passing; he places more emphasis on Luther's "more and more intense belief in divine providence," which for Luther required the "absolute acceptance of things as they were" (Ibid., p. 85).

21. Weber simply mentions Luther's experience with rebellious peasants as something which affected his view on worldly activity and cites in a footnote a letter of Luther's in support of his claim. Weber, note 19 of Chapter III, pp. 214–15.

22. Luther, *Works*, Volume 46, "The Christian in Society III," ed. Robert C. Schultz (Philadelphia: Fortress Press, 1967), p. 6.

In regard to my earlier discussion of enclosures and the loss of common rights in England (see Chapter 4, pp. 66–67), it is interesting to note that the fourth, fifth, and tenth articles of the Swabian peasantry were concerned with the elimination of, or interference with, rights such as those that were held by commoners in England. The fourth article concerns rights to game and fish; the fifth concerns the right to cut wood; and the tenth is concerned with the expropriation of meadows from the common fields. (The peasants' list of articles is reprinted in Luther, *Works*, Volume 46, pp. 8–16.) Luther declined to comment on any of these specific articles, which he thought the more proper province of lawyers. Luther, *Works*, Volume 46, p. 39.

23. Luther, *Works*, Volume 46, pp. 35–36.

24. Ibid., p. 49.

25. Ibid., p. 50.

26. Ibid., p. 49.

27. Weber indicates that Luther's final position on the calling is found in his lectures on Genesis (Weber, note 23 of Chapter III, p. 215). The editors of Luther's *Works* indicate in their "Introduction to

Volume 3" that Luther's lectures on Genesis were most likely written in the 1530s. Luther, *Works*, Volume 3, pp. ix–x.

28. Ibid., p. 62. See also p. 65.

29. Ibid., p. 65.

30. John (Jean) Calvin, *Institutes of the Christian Religion*, 3 vols., trans. Henry Beveridge (Edinburgh: Edinburgh Printing Company, 1845), Volume 2, p. 298.

31. Ibid., p. 299.

32. The emphatic sense of ambivalence is closely related to the Latin roots of the word: *ambo* = both; *valere* = to be strong. As a psychiatric term, it retained this sense of feeling strongly in two conflicting ways. That sense has been overshadowed by its more recent sense of indifference.

Elaine Pagels uses the term "deep ambivalence" to describe the conflicting teachings of early Christians in regard to sexuality, and she uses the word ambivalence in the same sense that I do. Pagels claims that this ambivalence "has resounded throughout Christian history for two millennia." Elaine Pagels, *Adam, Eve, and the Serpent* (New York: Random House, 1988), p. 29. Pagels limits her remarks to sexuality, however, while I focus on earthly activity other than sexual activity. Nevertheless, we are in agreement that Christianity is marked by this ambivalence.

Perhaps I should at this point offer some explanation for the exclusion of any detailed discussion of sexuality in my text. Although this facet of earthly delights is undoubtedly a crucial dimension of the modern subject, as Michel Foucault has so ably demonstrated, I am trying to uncover a different dimension of the self. To try to draw the arguments that I am going to make about the body and convenience close to Foucault's work on sexuality would only tend to obscure those features of the modern subject I am trying to highlight. I hope that I can get away, at least for now, with saying just that I make no pretense of presenting a complete or total perspective on the body in modernity; that was never my objective.

33. Luther, *Works*, Volume 1, p. 200.

34. Ibid.

35. Ibid., p. 198. Eve, of course, experienced fratricide in her family. Along with such violence, sixteenth-century mothers also encountered the danger of plagues.

36. Ibid., p. 202.

37. Ibid.

38. Ibid., p. 199.

39. Ibid., p. 201.

40. Ibid., p. 202.

41. Ibid., p. 206.

42. Ibid.

43. Ibid. See also pp. 208, 209, 216, for statements indicating that earthly life had become more difficult.

44. Ibid. See also p. 208: "The more closely the world approaches its end, the more it is overwhelmed by penalties and catastrophes."

45. Ibid., p. 204. See also p. 203.

46. Ibid., pp. 210–11. Luther also claimed that human diseases had increased as a result of man's growing sinfulness. See ibid., pp. 207–208.

47. Ibid., p. 209.

48. Ibid., p. 212. See also p. 204.

49. Ibid., p. 201. Luther is quoting Pindar.

50. Luther believed that the increase in the number and severity of punishments inflicted upon men was a sign that "everything will be destroyed, or Germany will pay the penalties for its sins in some other way." Luther, *Works*, Volume 2, p. 207. If the last day was not at hand, at least Germany was about to pay for its decadence, as had the cities of Sodom and Gomorrah (see p. 206).

51. Calvin, *Institutes*, Volume 2, p. 286. Augustine, in "De doctrina Christiana," offers a similar interpretation of life's miseries. Augustine describes how God maintains the health of "His body," the Church: "He disciplines it now and cleanses it with certain afflictions which act as medicines, so that, when it has been drawn forth from this world to eternity, He may join to Himself as His spouse 'the Church not having spot or wrinkle or any such thing.'" Augustine, *The Writings of St. Augustine*, Volume 2. pp. 38–39.

52. Calvin, *Institutes*, Volume 2, p. 286.

53. Ibid., p. 287.

54. Ibid., p. 294.

55. Ibid., pp. 293–94.

56. Calvin, *Sermon No. 10 on I Corinthians*, p. 698. Quoted in William J. Bouwsma, *John Calvin: A Sixteenth Century Portrait* (Oxford: Oxford University Press, 1988), p. 135.

57. Calvin, *Commentary on Psalms 104:31*. Quoted in Bouwsma, p. 135.

58. Calvin, *Institutes*, Volume 2, p. 295.

59. Ibid.

60. Ibid.

61. Weber acknowledges this in a footnote. Weber admits that these teachings of Calvin might in themselves "have opened the way to a very lax practice." See Weber, note 62 of Chapter IV, p. 272.

62. Although Calvin's attitude toward the sexual activity of married couples was quite tolerant, he does not focus on this form of pleasure in the tenth chapter of Book III of the *Institutes*, which is entitled "How to Use the Present Life and the Comforts of It." Bouwsma discusses Calvin's attitude toward sexuality and offers many citations from Calvin in support of his claim that Calvin held a positive attitude toward sexual activity. See Bouwsma, pp. 136–38.

63. Calvin, *Institutes*, Volume 2, pp. 295–96.

64. Ibid., p. 296.

65. Further indication of Calvin's ambivalence is found in his claim that "the contempt which believers should train themselves to feel for the present life, must not be of a kind to beget hatred of it..." Calvin, *Institutes*, Volume 2, p. 288.

66. Weber, p. 98.

67. Calvin, *Institutes*, Volume 2, p. 534.

68. In *The Protestant Ethic*, Weber mentions in a footnote that the first chapter of Ephesians is the principal biblical foundation for the doctrine of predestination (see Weber, note 15 of Chapter IV, p. 221). Calvin does rely, in part, on this chapter in his discussion of predestination. See, for example, Calvin, *Institutes*, Volume 2, pp. 543–45. There is some question as to the authenticity of this letter, however. For a discussion of the "deutero-Pauline" letters, see Pagels, pp. 23–25. But

Calvin also cites Paul's letter to the Romans, especially chapter 9, as a source of the doctrine (Calvin, *Institutes*, Volume 2, pp. 546–47), and there appears to be no question but that Paul did write this letter. So whether or not Ephesians was written by Paul, Calvin and Weber appear to be justified in citing Paul as a proponent of predestination.

69. In the ninth chapter of Romans, Paul also writes, "so then He has mercy on whom He desires, and He hardens whom He desires" (9:18). This verse, too, is cited by Calvin in support of the duality of predestination. Calvin, *Institutes*, Volume 2, pp. 544–45.

70. Calvin, *Institutes*, Volume 2, pp. 544–45.

71. As an example of predestination, Paul turns to the Old Testament account of God's choice between Jacob and Esau, the twin sons of Isaac and Rebekah (see Genesis 25:19–34). The Lord says to Rebekah, "Two nations are in your womb; and two peoples shall be separated from your body; and one people shall be stronger than the other; and the older shall serve the younger" (Genesis 25:23). In his letter to the Romans, Paul emphasizes that the fates of these twins were predestined by God according to his own will: "For though *the twins* were not yet born, and had not done anything good or bad, in order that God's purpose according to *His* choice might stand, not because of works, but because of Him who calls it was said to her, 'THE OLDER WILL SERVE THE YOUNGER'" (Romans 9:11–2). The choice of Jacob over Esau, which reversed the usual order of primogeniture, was made gratuitously, that is, without reference to the merit of Jacob or Esau. Similarly, claims Paul, God "has mercy on whom He desires, and He hardens whom He desires" (Romans 9:18).
In Calvin's elaboration of the doctrine of predestination, he expounds upon Paul's interpretation of this example from the Old Testament, and uses it to argue against the idea that salvation is based on merit or works. See Calvin, *Institutes*, Volume 2, pp. 546–48.

72. The doctrine of predestination became important for Augustine in the last decade of his long career in the service of the Roman Catholic Church. *The Predestination of Saints*, in fact, was one of Augustine's last books, written in 429, the year before he died. In his biography of Augustine, Peter Brown presents the doctrine of predestination as Augustine's answer to the barbarian invasion of the Roman Empire. Brown writes: "For Augustine's doctrine of predestination, as he elaborated it, was a doctrine for fighting men.... it was a doctrine of survival, a fierce insistence that God alone could provide men with an irreducible inner core" (Peter Brown, *Augustine of Hippo: A Biography*

[Los Angeles: University of California Press, 1970], pp. 403, 407). Such an inner core was needed among those who awaited the impending invasion of Africa by the Vandals and had to contemplate the possibility of their own martyrdom. Brown, p. 406.

But Augustine's doctrine of predestination appears to be more than just a response to the invasion of the Roman Empire. Augustine also used this doctrine, especially the element of gratuitous salvation, as a weapon against the remnants of the Pelagian heresy, which Augustine had fought in the second decade of the fifth century. In fact, Philip Schaff, in his *History of the Christian Church*, indicates that Augustine wrote *The Predestination of Saints* in response to the "semi-Pelagianism" which had emerged in Southern Gaul. Philip Schaff, *History of the Christian Church* (New York: Charles Scribner's Sons, 1886), Volume III, pp. 859–60.

Pelagius and his followers believed that men were able to attain salvation through the exertion of their own disciplined wills. See Pagels, pp. 124–25, 129, 130, 144; see also Brown, pp. 343, 346, 349, 350. Pagels emphasizes Augustine's interpretation of original sin, and the use to which he put it in his struggle with Pelagius and his followers, especially Julian. Pagels, pp. 127–50. She never mentions the doctrine of predestination, although Augustine, on more than one occasion, used this doctrine to counter Pelagian arguments.

For example, in *Admonition and Grace*, which Augustine wrote to settle some controversies that arose at the monastery at Hadrumetum, where some monks were affected by Pelagian ideas, Augustine emphasized that "we ought to understand that no one can be singled out of that lost mass for which Adam was responsible, except one who has this gift [of perseverance]; and he who has it, has it by the grace of the Savior." Augustine, *Admonition and Grace*, in Volume 4 of The Fathers of the Church Series (New York: Fathers of the Church, Inc., 1957), pp. 256–58. In this discussion, Augustine emphasizes that of those who are saved, "even those who have led the very worst kind of life are led to repentance through the goodness of God" (p. 258).

So in Calvin's emphasis on the gratuitousness of salvation, he seems to have firm support from the teachings of Augustine. Even though he did not make extensive use of those teachings in the *Institutes*, Calvin does point out after briefly mentioning Augustine's argument with the Pelagians concerning the gratuity of salvation, that "were we disposed to frame an entire Volume out of Augustine, it were easy to show the reader that I have no occasion to use any other words than his: but I am unwilling to burden him with a prolix statement" (Calvin, *Institutes*, Volume 2, p. 553). In this note, I have not been so kind.

73. Ibid., p. 236. See pp. 236–42 for Calvin's criticism of indulgences.

74. Ibid., p. 214.

75. Ibid., p. 545.

76. Thomas Aquinas, *Summa Theologica*, trans. Fathers of the English Dominican Province, Volumes 19–20 of The Great Books of the Western World Series, ed.-in-chief Robert Maynard Hutchins (Chicago: Encyclopedia Britannica, Inc., 1952), Volume 20, p. 859. For Aquinas' syllogistic reasoning on the necessity of the sacraments for salvation, see p. 855.

77. Calvin, *Institutes*, Volume 3, pp. 479–517.

78. Ibid., p. 313.

79. Ibid., pp. 315–16.

80. Weber, p. 102.

81. Ibid., pp. 104–105.

82. Weber points out in several places this feature of Lutheranism. See Weber, pp. 102, 126, note 104 of chapter IV, on p. 240.

83. Ibid., p. 106.

84. Ibid. See also p. 116. Weber portrays the Calvinist rejection of penitence as a harsh measure, one which eliminated a periodical catharsis of guilt for sin. According to Weber, the priest, as confessor, "dispensed atonement, hope of grace, certainty of forgiveness, and thereby granted release from that tremendous tension to which the Calvinist," because of the doctrine of predestination, "was doomed by an inexorable fate, admitting of no mitigation" (Weber, p. 117).

Although such tension may have been the consequence of Calvin's rejection of penitence, I should mention that Calvin's rejection was not intended to make things more difficult for members of the elect. And while Weber never explicitly claims such an intent for Calvin, his emphasis on the harsh consequence creates that impression. But in fact, Calvin criticized the demand for "full and complete" contrition, the first element of penitence, precisely because it produced anxiety, tension, and distress among sinners. (For a discussion of the tripartite formula of the sacrament of penance, see Calvin, *Institutes*, Volume 2, pp. 187–218.)

Priests, writes Calvin, "exact [contrition] as due, that is, full and complete: meanwhile, they decide not when one may feel secure of having performed this contrition in due measure.... when such bitterness of sorrow is demanded as may correspond to the magnitude of the offence, and be weighed in the balance with confidence of pardon, miserable consciences are sadly perplexed and tormented when they see that the contrition due for sin is laid upon them, and yet that they have no measure of what is due, so as to enable them to determine that they have made full payment" (Calvin, *Institutes*, Volume 2, p. 204).

Although Calvin recognized that the Catholic sacrament of penitence may have provided comfort to a "good part of the world," among the elect who were chosen for salvation, it was a source of anxiety. Those "who were affected with some sense of God" could never rest in confidence that they had been perfectly contrite, or had enumerated all of their sins. Calvin, *Institutes*, Volume 2, p. 205. So Calvin's rejection of penitence was not a harsh denial of comfort to believers, as Weber intimates, but was intended to provide comfort and confidence to the elect. Of course, Weber's larger and more important claim, which I will get to shortly, is that the Calvinist doctrine of predestination was itself a source of dread and anxiety. What I have tried to indicate in this note is that Calvin's rejection of penitence was not intended as such.

85. Aquinas, *Summa Theologica*, Volume 19 of The Great Books, p. 137.

86. Calvin's arguments can be found in *Institutes*, Volume 2, pp. 214–22. Again, although Calvin does not address Aquinas specifically, many of the distinctions Calvin attacks can be found in Aquinas's discussion of the sacrament of penitence (e.g. sins before and after baptism, mortal and venial sins). See Aquinas, *Summa Theologica*, Volume 20 of The Great Books, pp. 879–81.

87. Aquinas, *Summa Theologica*, Volume 19 of The Great Books, p. 134.

88. Calvin, *Institutes*, Volume 2, pp. 193–94.

89. Augustine, *Enchiridion* (also known as *Faith, Hope, and Charity*), trans. Bernard M. Peebles, in Volume 2 of The Fathers of the Church Series (Washington: The Catholic University of America Press, 1947), p. 449.

90. Augustine, *Against Julian*, trans. Matthew A. Schumacker, Volume 35 of The Fathers of the Church Series (New York: Fathers of the Church, Inc. 1957), p. 258.

91. Augustine, *Enchiridion*, p. 451. And again on p. 452, Augustine indicates that "all have been mingled together in one mass of perdi-

tion from a cause leading back to their origin." See also the quote from *Admonition and Grace* in note 72, in which Augustine mentions "the lost mass for which Adam was responsible."

92. The controversy at the monastery in Hadrumetum, which elicited Augustine's *Admonition and Grace* (see note 72), was based on this polemical interpretation of Augustine's doctrine of predestination. Some monks used that doctrine to question the practice of admonishing those monks who seemed to be weakening in their devotion to God. Since some were predestined to damnation, argued these recalcitrant monks, there was no point in punishing them or admonishing them. Augustine rejected this conclusion and replied: "Whenever you fail to follow the known commands of God and are unwilling to be admonished, you are for this very reason to be admonished, that you are unwilling to be admonished" (*Admonition and Grace*, pp. 250–51). For another example of Augustine's rejection of this mistaken inference from his doctrine, see *Enchiridion*, pp. 452–53.

93. Schaff, p. 856.

94. Ibid., pp. 863–65.

95. Ibid., p. 865.

96. Charles Joseph Hefele, *A History of the Councils of the Church, from the Original Documents*, trans. William R. Clark (Edinburgh: T & T Clark, 1895; New York: AMS Press, 1972), Volume IV, "A.D. 451–A.D. 680," p. 154.

97. Ibid., p. 165. The twenty-five canons, or Capitula, of the second Council of Orange, along with the creed, are printed in Hefele, pp. 155–65. For each of the canons, Hefele provides relevant citations of Augustine's writings.

98. Schaff, pp. 867, 879. Schaff notes that, despite the prevalence of what he calls "Semi-Augustinianism," there remained over the years proponents of a more stringent, strict Augustinianism, among whom were Wycliffe and Huss, "the precursors of the Reformation" (p. 870).

99. This reading of Augustine's doctrine of predestination, as including the saved and the damned, is certainly questionable. After all, this doctrine was the source of controversy within the Church for a century after Augustine's death. But I think it is a defensible reading. Philip Schaff, in his discussion of Augustine's conception of predestination (Schaff, Volume III, pp. 851–56) offers a seemingly opposite interpretation of the doctrine. Schaff claims that Augustine, "strictly speak-

ing, knows nothing of a *double* decree of election and reprobation, but recognizes simply a decree of election to salvation" (p. 853). And a little later, Schaff writes, "The predestination has reference only to good, not to evil" (p. 854). Schaff acknowledges, however, that Augustine does indeed mention the "predestination to perdition" in several places (pp. 854–55, note 1 on p. 869).

While this is not the place to elaborate an argument supporting the claim that Augustine held the idea that the damned, like the saved, were predestined, it should be pointed out that both sides in this argument agree that Augustine thought all men were damned since Adam's fall. Schaff, for example, agrees that according to Augustine, "once committed, [the fall] subjected the whole race, which was germinally in the loins of Adam, to the punitive justice of God" (Ibid., pp. 853–54).

100. Calvin, *Institutes*, Volume 2, pp. 533–4.

101. Ibid., p. 560.

102. Ibid., p. 561.

103. John Calvin, *Commentaries*, ed. and trans. Joseph Haroutunian, in collaboration with Louise Pettibone Smith, Volume xxiii of The Library of Christian Classics (Philadelphia: The Westminster Press, 1948), Chapter VII, "Election and Predestination," p. 298.

104. Calvin, *Institutes*, Volume 2, p. 561.

105. Ibid., p. 560.

106. This is the term Weber uses to describe the doctrine of predestination, in specific reference to Calvin. See Weber, p. 102.

107. Ibid. See also note 36 of Chapter IV, pp. 226–27.

108. Quoted in Weber, p. 100.

109. Ibid., p. 104.

110. Aquinas, *Summa Theologica*, Volume 20 of The Great Books, pp. 880–81.

111. Weber, p. 128.

112. Ibid., p. 110.

113. Calvin, *Institutes*, Volume 2, p. 290.

114. Ibid.

115. Weber, p. 110.

116. Ibid., p. 112.

117. Ibid., p. 114.

118. Ibid., pp. 114–15.

119. Ibid., p. 114. Calvin also claimed that "the special election which otherwise would remain hidden in God, he at length manifests by his calling." *Institutes*, Volume 2, p. 580. But in this context, Calvin is using calling in the sense of God's communication to man, and not in the sense of one's earthly occupation. In the same chapter, Calvin writes that "the nature and dispensation of calling...consists not merely of the preaching of the word, but also of the illumination of the Spirit" (p. 583).

120. Weber, p. 118.

121. Ibid., p. 111; note 41 of Chapter IV on p. 228; and note 48 of Chapter IV on p. 229.

122. Calvin, *Institutes*, Volume 2, pp. 530–31. See also p. 585.

123. Weber, p. 161. Weber notes the similarity between these Calvinist ideas and the contemporaneous, utilitarian arguments for the division of labor (p. 161) and insists that the latter were derived from the former. See note 33 of Chapter V, on pp. 265–66.

124. Ibid., p. 162.

125. Ibid.

126. Ibid., pp. 159–60.

127. Michael Walzer, *The Revolution of Saints: A Study in the Origins of Radical Politics* (Cambridge: Harvard University Press, 1965), pp. 217–19, 226–27. Although their arguments in general are quite distinct, Walzer does at one point agree with the claim made by Weber that the Puritans "in some fashion mediated" the transition from traditional to modern society (Walzer, p. 230). In the sixth chapter of *The Revolution of Saints*, entitled "The New World of Discipline and Work," Walzer indicates that the discipline of the Puritans was not directed solely at themselves, but was also focused on those selves which had not yet assumed the form of the disciplined subject of modernity.

128. Weber indicates that Lutherans frequently charged Calvinists with returning to the Catholic idea of salvation by works. See Weber, p. 115.

129. Ibid., pp. 156–57.

130. Ibid., p. 157.

131. Ibid.

132. Ibid.

133. Ibid., note 14 of Chapter V on p. 261. In this footnote, Weber acknowledges that in regard to the regulation of the way in which time was spent, "Protestant asceticism follows a well-beaten track" which began in the monasteries. Jacques LeGoff, in *Time, Work & Culture in the Middle Ages*, makes a similar point. In an essay entitled "Labor Time in the 'Crisis' of the Fourteenth Century," LeGoff points out that in the regulation of time, "the Church took initiatives. Monks, especially...were masters in the use of *schedules*" (Jacques LeGoff, *Time, Work & Culture in the Middle Ages*, trans. Arthur Goldhammer [Chicago: The University of Chicago Press, 1980], p. 48).

134. LeGoff makes the case that sometime around the fourteenth century, Christian merchants began to develop a "temporal horizon" which differed from that of the Church. The episodic, cyclical sense of time held by the Church and institutionalized in the ringing of bells for the canonical offices was unable to provide the precision and regularity which were required by merchants (LeGoff, pp. 35–36). It was in the fourteenth century, partly in response to this mercantilistic need and due in part to the development of accurate mechanical clocks, that the day was divided into twenty-four equal parts (Ibid., pp. 48–49). LeGoff also points out that the late medieval merchant used this measurable sense of time not only to regulate his business activity and labor force, but that he also "introduced his business organization into everyday life and regulated his conduct according to a schedule." For LeGoff, this marks "a significant secularization of the monastic manner of regulating the use of time" (Ibid., p. 51).

It would seem, therefore, that the Puritans were not the first to regulate time outside of the monastery. I think Weber would have acknowledged that these merchants of the late Middle Ages were part of that "well-beaten track" which the Puritans followed. But the difference between the medieval merchant and the Puritan entrepreneur is that, for the latter, the regulation of his time was part of an ascetic practice, whereas for the former, part of his schedule included "the time of rest...diversion, and visiting, the leisure and social life of men and substance" (Ibid., p. 51). As will be indicated immediately following in the text, Puritan asceticism viewed such leisure and diversion as a waste of time and urged that such activities be kept to a minimum.

Because I will have no other opportunity in this essay for considering LeGoff's writings, I must mention one more of his insights—a particularly fascinating one from the perspective I have been developing. In an essay entitled "Merchant's Time and Church's Time in the Middle Ages," LeGoff relates the development of late medieval mercantilism with its discontinuous, highly measurable temporal horizon and the increased importance of the Catholic sacrament of penance. LeGoff argues that one of "the principal criticisms leveled against the merchants" by the Church "was the charge that their profit implied a mortgage on time, which was supposed to belong to God alone" (Ibid., p. 29). The merchant's profits depended to a certain extent on his taking advantage of time. He exacted interest from those who could not settle their account at the moment, and he took advantage of the fluctuation over time of prices for goods. Because of this profitable use of time, which belonged to God, not man, the merchant could be interpreted as "committing usury *by selling what does not belong to him*" (Ibid).

LeGoff further argues that, in response to this threat posed to the Church by the development of mercantilism, the Church not only elaborated a "theologico-moral theory of usury," but it also deployed the sacrament of confession to help assimilate the merchant and his new sense of time (Ibid., p. 38). Around the twelfth century, claims LeGoff, penance shifted from "external sanction to internal contrition." Confession became a way of discovering "internal dispositions to sin and redemption, dispositions rooted in concrete social and professional situations" (Ibid., p. 39). The merchant had to become conscious of his intentions and dispositions in that temporal activity which threatened to develop outside the purview of the Church. From LeGoff's perspective, the sacrament of confession—which, it will be recalled, became institutionalized by the Church in the thirteenth century (see p. 135 of this chapter)—helped to close "the merchant's loophole" and reunite "the time of salvation and the time of business" (Ibid.).

135. Weber, p. 166.

136. Quoted by Weber, in note 14 of Chapter V, p. 261.

137. Ibid., p. 171.

138. Ibid., p. 172.

139. Ibid., p. 181.

140. Ibid.

141. Ibid. This is the only other reference to the cage Weber makes in his text, and it immediately follows the long quote cited on p. 145 of

this chapter. The complete sentence reads: "To-day the spirit of religious asceticism—whether finally, who knows?—has escaped from the cage."

CHAPTER 7. NIETZSCHE AND MODERN ASCETICISM

1. Max Weber, *The Protestant Ethic and the Spirit of Capitalism*, trans. Talcott Parsons (London: Unwin Hyman, 1930), p. 183.

2. Friedrich Nietzsche, *The Will to Power*, ed. Walter Kaufmann, trans. Kaufmann and R. J. Hollingdale (New York: Vintage Books, 1968), note 88, p. 54.

3. I am aware that Nietzsche recognized as a "great lie in history" the idea that the Reformation was caused by the corruption of the Church. See Nietzsche, *The Will to Power*, note 381, p. 205. My claim, however, is not directed at the corruption of the Church—for example, its practice of selling indulgences. My point here is not so much that the Church was corrupt, but that sacramental developments, the relaxation of the prohibition of usury, and the revaluation of mercantilism all tended to weaken people's concern with the afterlife and their religion.

4. Nietzsche, *The Will to Power*, note 89, pp. 54–55.

5. *Compact Edition of the Oxford English Dictionary*, Volume 1, p. 1324.

6. Friedrich Nietzsche, *The Genealogy of Morals: An Attack*, trans. Francis Golffing (Garden City, New York: Doubleday & Company, 1956), Essay III, Chapter i, p. 231.

7. Ibid., III, xiii, 256.

8. Ibid., III, xv, 262.

9. Ibid., III, xvii, 266.

10. Ibid., III, xv, 263.

11. Ibid., III, xv, 267.

12. Ibid., III, xv, 263.

13. Ibid.

14. Ibid., III, xxi, 279.

15. Ibid., III, xvii, 268.

16. Ibid., III, viii, 243.

17. Nietzsche, *The Will to Power*, note 223, p. 130.

18. Ibid.

19. In the third essay of *The Genealogy of Morals*, in which he examines asceticism in detail, Nietzsche is not a constant foe of asceticism. In regard to the asceticism of philosophers, Nietzsche describes it in an affirmative manner and with more than a bit of sympathy (see Chapters vii–ix). He admits "that a certain asceticism, that is to say a strict yet high-spirited continence, is among the necessary conditions of strenuous intellectual activity as well as one of its natural consequences" (III, ix, 247). And in regard to those "mighty slogans"—poverty, humility, and chastity—he says that "when we examine the lives of the great productive spirits closely, we are bound to find all three present in some degree" (III, viii, 243).

20. Nietzsche, *The Will to Power*, note 916, pp. 483–84.

21. Ibid., note 226, p. 131. I am aware of the criticism that could be leveled against such an interpretation of Christian asceticism. For example, Peter Brown claims that those who see in the ascetic tradition of the "desert fathers" nothing more than contempt and hatred for the body "miss its most novel and its most poignant aspect" (Peter Brown, *The Body and Society: Men, Women and Sexual Renunciation in Early Christianity* [New York: Columbia University Press, 1988], p. 235). The aspect Brown has in mind here is that, for those desert ascetics "the body was not an irrelevant part of the human person, that could, as it were, be 'put in brackets.'... It was, rather, grippingly present to the monk" (p. 236).

Although Brown does not attack Nietzsche directly, and this is probably not the place to attempt a defense of Nietzsche's interpretation of Christianity's attitude toward the body, I would like to say simply that, despite the variety of attitudes toward the body which can be identified in the many strands of early Christianity, there is, nevertheless, a certain amount of contempt for the mortal, fallen body in Christian thought in general. It is true, as Brown emphasizes, that the desert fathers disciplined their bodies in order to transform their hearts and had no thought of abandoning their bodies. But these ascetics nevertheless sought to be rid of the corrupt, fallen, mortal body, as Brown also indicates:

> In its 'natural' state—a state with which the ascetics tended to identify the bodies of Adam and Eve—the body acted like a finely tuned engine, capable of 'idling' indefinitely. It was only the twisted will of

> fallen men that had crammed the body with unnecessary food, thereby generating in it the dire surplus of energy that showed itself in physical appetite, in anger, and in the sexual urge. In reducing the intake to which he had become accustomed, the ascetic slowly remade his body.... Its drastic physical changes, after years of ascetic discipline, registered with satisfying precision the essential, preliminary stages of the long return of the human person, body and soul together, to an original, natural and uncorrupted state" (p. 223).

It was in this natural, uncorrupted state, free from the urges and appetites of the mortal body, that the desert fathers appreciated the body. Nietzsche's point is that, in this ideal state, the body is a "cadaverous abortion" (Nietzsche, *The Will to Power*, note 226, p. 131) and that Christianity in general was uncomfortable, to say the least, with the mortal, "corrupted" body.

22. Ibid., note 227, p. 131.

23. Ibid., note 224, p. 130.

24. Nietzsche, *The Genealogy of Morals*, I, ii, 232.

25. Ibid.

26. Ibid., I, ii, 233.

27. Ibid.

28. Ibid., III, xviii, 271.

29. Ibid., III, xxiii, 284.

30. Ibid.

31. Ibid., III, xxv, 290.

32. Ibid., III, xxiv, 288.

33. Weber, p. 48.

34. Weber points out in several places that religion did not play a significant role in Franklin's thoughts about worldly activity. See Weber, p. 180; and note 36 on p. 227. See also p. 48. From Franklin's correspondence, however, it appears that, although Franklin may have abandoned writing about religious issues early in his career (see Franklin's letter to Benjamin Vaughn, reproduced in part in Benjamin Franklin, *The Autobiography and Other Writings*, ed. L. Jesse Lemisch [New York: Signet Classics, 1961], pp. 328–29), he did offer some reflections on religion late in his life. I will mention these in the text that follows.

35. Weber, p. 53.

36. Benjamin Franklin, *The Complete Works of Benjamin Franklin*, ed. John Bigelow (New York: G. P. Putnam's Sons, 1887), Volume II, p. 118.

37. Franklin, *The Autobiography*, pp. 189–90. Weber does not cite this later essay, which was written as an introduction to the twenty-fifth edition of Poor Richard's Almanack, "reprinted in hundreds of editions...and was translated into French, German, Italian, and more than a half-dozen other languages" (p. 188).

38. Ibid., p. 193.

39. Ibid., p. 194.

40. Ibid., p. 195.

41. Ibid., pp. 217–18.

42. Ibid., p. 218. The letter was to Cadwallader Colden, a long-time friend of Franklin's, who, like Franklin, was interested in science and politics.

43. Ibid. This letter was to John Fothergill, a Scottish doctor.

44. Ibid., pp. 239–43.

45. Ibid., p. 241.

46. Ibid., p. 243.

47. Ibid., p. 237.

48. Ibid.

49. Ibid., p. 223.

50. Franklin's description of this device is quoted in Carl van Doren, *Benjamin Franklin* (New York: The Viking Press, 1938), pp. 168–69.

51. Seymour Dunbar, *A History of Travel in America* (Indianapolis: The Bobbs-Merrill Company, 1915), pp. 45–46.

52. Quoted in van doren, p. 741.

53. Franklin, p. 329.

54. Ibid., p. 337.

55. Ibid., p. 239; and van Doren, p. 173.

CHAPTER 8. TRACES OF MODERN ASCETICISM

1. Thomas Hobbes, *Leviathan, or The Matter, Forme & Power of a Common-Wealth Ecclesiasticall and Civill* (Oxford: Oxford University Press, 1909).

2. I tend to side with those who recognize an ambiguity in regard to the role of Christianity in Hobbes's thought. Benjamin Milner makes a case for such ambiguity in "Hobbes on Religion," *Political Theory* 16, 3, 400–25. While I agree with the overall thrust of Milner's argument—that Hobbes "has attempted the synthesis of a natural and a biblical theology, and that it is the imperfect assimilation of these that leaves his philosophy open to such antagonistic interpretations" (p. 400)—I am not making a similar argument. I do not want to stress the way in which Hobbes tried to assimilate these two different theologies. What I want to stress is the manner in which Hobbes broke from traditional, scriptural Christianity. The difference between the two theologies Milner identifies is my concern.

And in regard to Milner's choice of "ambiguity" to describe the tension in Hobbes's thought, I agree. This tension cannot be described as ambivalence, the term I used to describe the different tension I identified in the thought of Luther and Calvin (see Chapter 6, pp. 123–129.) In the case of these reformers, there was a strong feeling in two opposing directions—the love, and contempt, for earthly life. In Hobbes, however, there is not a strong feeling for scriptural theology. I agree with Milner that one of Hobbes's purposes in *Leviathan*—Milner calls it the "deepest intention" (p. 400)—is to control religion, and to put Christianity in the service of the commonwealth. An ambiguity in Hobbes's thought results from this attempt to render Christianity as a constant and consistent support of civil authority. But there is no ambivalence in Hobbes as concerns the independent value of Christianity as a theological system. The value of Christianity for Hobbes lies in the use to which it can be put in establishing and maintaining civility, order. This will become apparent in my argument that follows.

3. Hobbes, p. 99.

4. Ibid., pp. 94–95.

5. Ibid., pp. 95–96.

6. Leo Strauss, *The Political Philosophy of Hobbes: Its Basis and Genesis*, trans. Elsa M. Sinclair (Chicago: The University of Chicago Press, 1952), Chapter 2, "The Moral Basis," pp. 6–29. In this chapter, Strauss argues that, for all the attention paid to Hobbes's scientific

approach to politics, his political philosophy is ultimately grounded in a "moral and humanist antithesis of fundamentally unjust vanity and fundamentally just fear of violent death" (p. 27). I appreciate Strauss's elaboration of the role which the fear of death plays in Hobbes's thought, and I will further develop this theme in my argument. The passion of vanity, however, is much more important for Strauss's argument than it is for mine, so nothing more need be said here about this particular passion.

7. Hobbes, p. 96.

8. Ibid. As Hobbes puts it, "For WARRE, consisteth not in Battell onely, or the act of fighting; but in a tract of time, wherein the Will to contend by Battell is sufficiently known."

9. Ibid., p. 97

10. Ibid., p. 100.

11. Ibid., p. 98.

12. Ibid., p. 99.

13. Ibid.

14. Ibid., p. 100.

15. Ibid.

16. Ibid., p. 98.

17. Ibid., p. 97.

18. Strauss, pp. 20–21. Strauss continues his description of the struggle to the point where a master-slave relationship emerges, in which one of the combatants, for fear of his life, submits to the other. Strauss's account of this development is helpful not only for understanding Hobbes's political philosophy, but also for understanding Hegel's phenomenology of self-consciousness, which hinges on the master-bondsman relationship. For now, let me just say that, in Strauss's reading, Hobbes seems to share with Hegel a keen awareness of the role death plays in self-consciousness.

19. Hobbes, p. 101.

20. Ibid., pp. 101–102.

21. Ibid., pp. 138, 152–54.

22. Ibid., pp. 241–42.

23. Ibid., p. 242.

24. Ibid., p. 163.

25. Ibid., p. 167.

26. Ibid., p. 238. Earlier in the text, in his discussion of the nature of a valid covenant, Hobbes claims that "a Covenant not to defend my selfe from force, by force, is always voyd...no man can transferre, or lay down his Right to save himselfe from Death, Wounds, and Imprisonment." Ibid., p. 107. See also pp. 101–102

27. Ibid., pp. 107–108.

28. Ibid., p. 102.

29. Ibid., p. 98. See also p. 128, where Hobbes identifies the "finall Cause, End, or Designe of men" who choose to restrict their liberty in order to live in a commonwealth. This cause is "the foresight of their own preservation, *and of a more contented life thereby*" (my emphasis).

30. Hobbes's choice of the word *commodious* to describe the sort of life individuals hope to live in a commonwealth is an interesting one in light of the etymology of *convenience* I offered at the end of the second chapter. Back then, I pointed out a shift in the meaning of *convenience* that appears to have occurred sometime around the sixteenth century. Prior to this period, *convenience* meant "appropriate" or "suited" to something like a moral order or nature. After this shift, *convenience* referred back to the self and came to mean personal comfort or ease, or "commodiousness." In choosing *commodious* to describe life in a commonwealth, Hobbes seems to be invoking the modern sense of convenience as personal ease. But Hobbes had not quite made that shift in regard to the use of convenience itself. In the sentence immediately following the one quoted in my text, in which the idea of commodious living appears, Hobbes writes, "And Reason suggesteth convenient Articles of Peace, upon which men may be drawn to agreement" (p. 98). It is obvious that Hobbes, although writing in the seventeenth century, is here using *convenient* in its premodern sense of being appropriate to the rational order of nature, not being conducive to personal ease. Nevertheless, the rational laws that Hobbes finds in nature do indeed promote convenience in the modern sense of the word. One does not have to wait until Franklin, however, to find evidence for the etymological shift in convenience. As I will soon indicate, Locke, writing later in the seventeenth century than Hobbes, had made

that shift.

31. Hobbes, pp. 457–58.

32. Ibid., p. 457.

33. Ibid., p. 458.

34. Ibid.

35. Ibid., p. 462. For Hobbes's scriptural support for this claim, see pp. 462–66.

36. Ibid., p. 459.

37. Ibid., p. 470.

38. Ibid.

39. Ibid., pp. 470–71.

40. Ibid., p., 356. Milner claims in his article "Hobbes and Religion" (see note 2) that, to the best of his knowledge, Hobbes never mentions the fallen nature of man. Hobbes does mention it, but the concept of the fall does not really play a significant role in his thought.

41. Thomas Pangle, in *The Spirit of Modern Republicanism: The Moral Vision of the American Founders and the Philosophy of Locke* (Chicago: The University of Chicago Press, 1988), points out that "Locke devoted a very substantial portion of his published (and unpublished) writings to the project of forging and promulgating a new, 'reasonable' Christian theology" (p. 151). In the third part of this book, Pangle examines this Lockean project and concludes that "the truth or even the rational plausibility of the key articles of faith—the very articles that are crucial from the point of view of the usefulness of religion—Locke promises over and over to show without ever redeeming his pledge" (p. 214). This blatant failure on Locke's part leads Pangle to reconsider the importance which Locke placed on Christianity as a foundation for rational, moral activity (p. 201) and to "wonder whether Locke does not in fact look forward to, and attempt to help foster, a world where the educated classes would talk less and less of the afterlife and hence less and less of 'natural law'" (p. 215).

42. John Locke, *The Reasonableness of Christianity, as Delivered in the Scriptures* (Boston: T. B. Wait and Company, 1811), p. 1.

43. Ibid.

44. It is interesting to note that Locke, in the interpretation of the fall which opens *The Reasonableness of Christianity*, does not specify

that the forbidden tree was the tree of knowledge of good and evil. Perhaps Locke sensed a similarity between the hubris that led Adam and Eve to seek such knowledge, and his own rational impulses.

45. Ibid., p. 3.

46. Ibid.

47. Even though Locke explicitly rejects the idea that the state of nature is a state of war (see John Locke, *Two Treatises of Government* [London: J. M. Dent & Sons Ltd., 1924], Treatise II, Secs. 16–20, pp. 125–27.), he nonetheless agrees with Hobbes that the state of nature was a dangerous place. "If man in the state of Nature be so free..." asks Locke, "why will he part with his freedom, this empire, and subject himself to the dominion and control of any other power? To which it is the obvious answer, that though in the state of nature he hath such a right, yet the enjoyment of it is very uncertain and constantly exposed to the invasion of others; for all being kings as much as he, every man his equal, and the greater part no strict observers of equity and justice, the enjoyment of the property he has in this state is very unsafe, very insecure" (*Two Treatises*, II, 123, 179). Individuals unite to form civil society, therefore, in order to protect their property, by which Locke means not just chattel property, but their lives and liberties as well. People form a civil society "for the mutual preservation of their lives, liberties, and estates," writes Locke (*Two Treatises*, II, 123, 180).

48. Locke, *The Reasonableness of Christianity*, p. 6.

49. Locke writes, "I easily grant that civil government is the proper remedy for the inconveniences of the state of Nature." Locke, *Two Treatises*, II, 13, 123. See also II, 90, 160; and II, 127, 180.

50. In each of the three citations in the preceding note, Locke identifies as the source of the inconveniences of the state of nature the condition that each person has the right to be his or her own judge in any conflict in that state. Individuals are equal in their ability to exercise this natural right to punish offenses against the laws of nature.

In regard to Hobbes's recognition of natural liberty and equality as the source of the incommodities of the state of nature, see pp. 169–170.

51. For Augustine's view that all were damned through Adam's sin, see Chapter 6 of this text, pp. 135–136. For Luther's teaching that the punishments which were inflicted upon Adam were multiplied and increased over the years, see Chapter 6, pp. 124–126. For Luther and Calvin, it should be further recalled, the idea that all were punished for

Adam's sin was one pole of their ambivalent stance to earthly life.

52. Locke, *The Reasonableness of Christianity*, pp. 6, 9.

53. Ibid., pp. 1–2.

54. Ibid., p.7. See also pp. 2–4.

55. Ibid., p. 7.

56. Ibid.

57. A similar reading of Locke's perspective on the fall is presented in Thomas Pangle's *The Spirit of Modern Republicanism*, pp. 144–47. Pangle summarizes Locke's teachings on the fall as follows: "As a result of what Adam did, what each and every one of the rest of us was 'exposed to' was not punishment, but only the loss of the possibility of eternal life and immortal happiness, and the acquisition in their place of an existence filled from the outset with the 'toil, anxiety, and frailties of this mortal life...'" (p. 146).

58. Locke's argument here, however, is not based on the definition of punishment as it was in *The Reasonableness of Christianity*, but on the rules of grammar. Locke argues that Filmer's interpretation of Genesis 3:16 requires that the usual rules of grammar be inverted. According to Locke, when God speaks to Adam in the singular, Filmer reads this to mean mankind in general. This is the only way patriarchal authority for Adam's posterity can be derived from this passage in scripture, claims Locke. Locke, *Two Treatises on Government*, I, 46, 32–33.

59. Ibid., I, 47, 33.

60. Ibid.

61. Ibid.

62. Hannah Arendt, *The Human Condition* (Chicago: The University of Chicago Press, 1958), pp. 315–16.

63. Karl Marx and Friedrich Engels, *The Marx-Engels Reader*, second edition, ed. Robert C. Tucker (New York: Norton & Company, Inc., 1978), pp. 70–81.

64. Ibid., p. 76.

65. Ibid., p. 75.

66. Ibid.

67. Ibid., p. 77.

68. Ibid., p. 91.

69. Karl Marx and Friedrich Engels, *The German Ideology*, ed. R. Pascal (New York: International Publishers, Inc., 1939), pp. 67–68.

70. Ibid., p. 66.

71. Karl Marx, *Grundrisse: Foundations of the Critique of Political Economy*, trans. Martin Nicolaus (New York: Vintage Books, 1973), p. 325.

72. Ibid. The final clause of this quote reads: "because a historically created need has taken the place of a natural one." It is because of this substitution of historical for natural need that natural necessity disappeared. This contrast between historical and natural needs also appeared in *The German Ideology*, where Marx and Engels wrote, "as soon as a need is satisfied, (which implies the action of satisfying, and the acquisition of an instrument), new needs are made; and this production of new needs is the first historical act" (pp.16–17). For Marx, therefore, the historicity of human being, as opposed to animal life-activity, is based on the development of new needs. It will soon be indicated in the text below, how much Marx appreciated the development of needs under capitalism.

73. Marx and Engels, *The Marx-Engels Reader*, p. 89.

74. Marx, *Grundrisse*, p. 409.

75. Karl Marx, *Capital: A Critique of Political Economy*, Volume 3, trans. David Fernbach (New York: Vintage Books, 1981), pp. 958–59. The particular passage in *Capital* from which this quote is taken, and upon which I will focus, is also found in Tucker's *Marx-Engels Reader*, pp. 439–41, in a different translation.

76. Ibid., p. 959.

77. Ibid.

78. Ibid.

79. Marx emphasizes this point in the third Volume of *Capital*: "But this always remains a realm of necessity" (p. 959).

80. Edmond Preteceille and Jean-Pierre Terrail, *Capitalism, Consumption and Needs,* trans. Sarah Matthews (Oxford: Bsisl Blackwell, 1985), p. 176.

81. Ibid., p. 99.

82. Ibid.

83. Ibid., p. 198.

84. Ernest Mandel, *Late Capilalism,* trans. Joris DeBres (London: NLB, 1975; original German edition, 1972), p. 395

85. Ibid., p. 397.

86. Ibid., pp. 394–95.

87. Ibid., p. 395.

88. Terrail uses these latter terms, which he borrows from Marx's discussion of precapitalist economic formations, to describe the point of view which could recognize in the ancient world any superiority to modernity, with the latter's "unrestricted development of the capacities and needs of the producers" (Preteceille and Terrail, pp. 49–50). Although my argument has not taken the shape of a contest between antiquity and modernity, nor has it been cast in terms of a hierarchical relation which elevates real, physical needs over developed, refined ones, I think Preteceille and Terrail would nevertheless take my critique of modern consumption practices to be of as little value as such arguments.

CHAPTER 9. THE END OF DEATH

1. Martin Heidegger, *Being and Time*, trans. John Macquarrie and Edward Robinson (New York: Harper & Row, 1962).

2. Ibid., p. 27, note 1.

3. Ibid., p. 280. Heidegger claims that "the reason for the impossibility of experiencing Dasein ontically as a whole which is, and therefore of determining its character ontologically in its Being-a-whole, does not lie in any imperfection of our *cognitive powers*. The hindrance lies rather in the *Being* of this entity."

4. Ibid., p. 279.

5. Ibid., pp. 280–81.

6. Ibid., p. 282.

7. Ibid., p. 294.

8. Ibid., p. 295. See also p. 298.

9. Ibid., p. 295.

10. Ibid., p. 297.

11. Ibid., p. 298.

12. Ibid., p. 307.

13. Ibid., p. 298.

14. Ibid., p. 305.

15. Ibid., p. 307.

16. See ibid., p. 292, for Heidegger's comments on otherworldly immortality.

17. Ibid., p. 308.

18. Ibid., p. 311.

19. Ibid., p. 310.

20. In *Eros and Civilization*, Marcuse employs this dichotomy when he claims that "the more external to the individual the necessary labour becomes, the less does it involve him in the realm of necessity. Relieved from the requirements of domination, the quantitative reduction in labor time and energy leads to a qualitative change in human existence: the free rather than the labor time determines its content" (Herbert Marcuse, *Eros and Civilization: A Philosophical Inquiry Into Freud* [New York: Vintage Books, 1962; originally published Boston: Beacon Press, 1955], pp. 203–204).

21. Herbert Marcuse, *An Essay on Liberation* (Boston: Beacon Press, 1969), p. 4.

22. Ibid.

23. Marcuse, *Eros and Civilization*, p. 204.

24. Marcuse, *An Essay on Liberation*, Chapter 2, "The New Sensibility," pp. 23–48. See also Herbert Marcuse, *One-Dimensional Man: Studies in the Ideology of Advanced Industrial Society* (Boston: Beacon Press, 1964), pp. 230–31, where Marcuse claims:

> Within the established societies, the continued application of scientific rationality would have reached a terminal point with the mechanization of all socially necessary but individually repressive labor ("socially necessary" here includes all performances which can be exercised more effectively by machines, even if these performances produce luxury and waste rather than necessities). But this stage would also be the

> end and limit of the scientific rationality in its established structure and direction. Further progress would mean the *break,* the turn of quantity into quality. It would open the possibility of an essentially new human reality—namely, existence in free time on the basis of fulfilled vital needs. Under such conditions, the scientific project itself would be free for trans-utilitarian ends, and free for the 'art of living' beyond the necessities and luxuries of domination. In other words, the completion of the technological reality would be not only the prerequisite, but also the rationale for *transcending* the technological reality.

Even though science and technology do not appear to have become "free for the 'art of living,'" at least Marcuse hoped for a break with technical reality.

25. Marcuse, *Eros and Civilization*, p. 211.

26. Ibid.

27. Ibid.

28. Herbert Marcuse, "The Ideology of Death," in Herman Feifel, ed., *The Meaning of Death* (New York: McGraw-Hill Book Company, 1959), p. 66.

29. Ibid., p. 69.

30. Ibid.

31. Ibid., p. 67.

32. Ibid., p. 66.

33. Marcuse, *Eros and Civilization*, p. 216.

34. Marcuse, "The Ideology of Death," p. 73.

35. Marcuse, "The Ideology of Death," p. 69.

36. Marcuse, *Eros and Civilization*, p. 215.

37. Michel Foucault, *The History of Sexuality*, Volume 1, trans. Robert Hurley (New York: Vintage Books, 1980), p. 136.

38. Ibid., p. 137.

39. Robert Jastrow, *The Enchanted Loom: Mind in the Universe* (New York: Simon and Schuster, 1981), pp. 166–7. I first read this passage in an article by David Lavery, entitled "Departure of the Body Snatchers, or the Confessions of a Carbon Chauvinist," *The Hudson Review*, Autumn, 1986. In this article, Lavery recounts his encounter with a body-snatcher during a panel discussion on "Computers,

Robots, and You." Faced with Lavery's reluctance to abandon his body in order to live in "indestructible lattices of silicon," the body-snatcher responded by calling Lavery a carbon chauvinist.

40. Foucault, *The History of Sexuality*, Volume 1, p. 138.

41. *Askesis*, the Greek root of the English *ascesis*, means exercise. See Oxford Dictionary of English Etymology. Michel Foucault discusses the relation between Greek moral askesis and the training of the body in *The Use of Pleasure*, Volume 2 of The History of Sexuality, trans. Robert Hurley (New York: Pantheon Books, 1988), pp. 72–77.

42. See Peter Brown, *The Body and Society: Men, Women and Sexual Renunciation in Early Christianity* (New York: Columbia University Press, 1988), pp. 213–17, for a discussion of the ascetic opportunities which the desert provided for Christians in Egypt.

It is interesting to note that Jean Baudrillard identifies an American form of asceticism which has much to do with the desert. Baudrillard writes: "America always gives me a feeling of real asceticism. Culture, politics—and sexuality too—are seen exclusively in terms of the desert, which here assumes the status of a primal scene. Everything disappears before that desert vision. Even the body, by an ensuing effect of undernourishment, takes on a transparent form, a lightness near to complete disappearance" (Jean Baudrillard, *America*, trans. Chris Turner [London: Verso, 1988, originally published in French in 1986], p. 28). I would like to suggest that the visions of extraterrestrial colonization and star travel serve an ascetic function in modernity much like that which the desert served for Christianity.

INDEX